'A visionary, practical and lyrical book on restoring land, from one of the best in the game, on the front line of nature restoration.'
Benedict Macdonald, author of *Rebirding*

'*Wild Fell* is a beautiful, powerful book that subtly navigates great and complex challenges.'
George Monbiot, author of *Regenesis*

'In a country defined as the seventh most nature depleted on Earth, in a region plagued by flooding and climate-chaos, here comes Lee Schofield's brilliant book full of positive action and hope for the future. *Wild Fell* throws down a gauntlet to us all to make the Lake District a national park that is genuinely worthy of the title.'
Mark Cocker, author of *Our Place*

'A soaring elegy to nature, a book infused with a deep love of place, and a stirring call to restore wildlife to our landscapes, written with wit, verve and humility.'
Guy Shrubsole, author of *The Lost Rainforests of Britain*

'As the competing needs of agriculture and conservation jostle for ascendency, land management in Britain has reached a tipping point. Candid, raw and searingly honest, Lee Schofield offers a naturalist's perspective of the challenges unfolding in the ancient yet ever-changing landscape of Haweswater and shares with us his gloriously vibrant vision for the future.'
Katharine Norbury, author of *The Fish Ladder*

'Exhilarating . . . His writing, like the extinct, extant and envisioned landscapes he describes, is studded with moments of immense beauty – you can almost smell rock and moss and nectar, hear butterflies and grasshoppers flit and whirr, feel the shadow of a great wing passing between you and the sun.'
British Wildlife

Wild Fell

Fighting for Nature on a
Lake District Hill Farm

Lee Schofield

PENGUIN BOOKS

TRANSWORLD PUBLISHERS
Penguin Random House, One Embassy Gardens,
8 Viaduct Gardens, London sw11 7bw
www.penguin.co.uk

Transworld is part of the Penguin Random House group of companies
whose addresses can be found at global.penguinrandomhouse.com

First published in Great Britain in 2022 by Doubleday
an imprint of Transworld Publishers
Penguin paperback edition published 2023

Map by Liane Payne
Illustrations by Bob Venables

A CIP catalogue record for this book
is available from the British Library.

ISBN
9781804990964

Typeset in Granjon LT Std by Jouve (UK), Milton Keynes.
Printed and bound in Great Britain by Clays Ltd, Elcograf S.p.A.

The authorized representative in the EEA is Penguin Random House Ireland,
Morrison Chambers, 32 Nassau Street, Dublin d02 yh68.

Penguin Random House is committed to a sustainable
future for our business, our readers and our planet. This book
is made from Forest Stewardship Council® certified paper.

For Elliot and Aphra

CONTENTS

Wild Fell

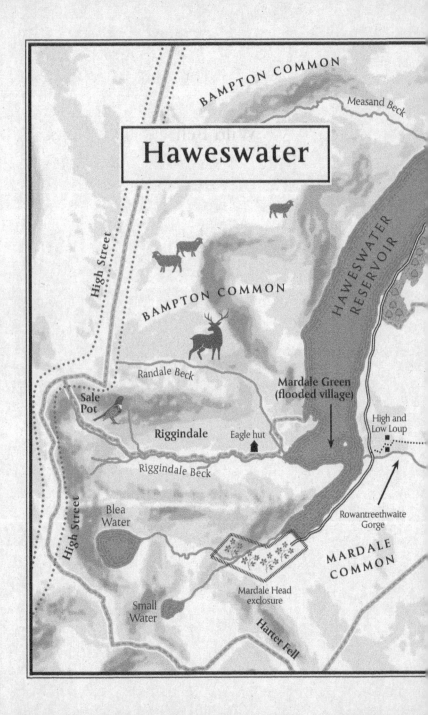

Haweswater

BAMPTON COMMON

Measand Beck

HAWESWATER RESERVOIR

High Street

BAMPTON COMMON

Randale Beck

Sale Pot

Riggindale

Eagle hut

Mardale Green (flooded village)

High and Low Loup

Rowantreethwaite Gorge

Riggindale Beck

High Street

Blea Water

Mardale Head exclosure

MARDALE COMMON

Small Water

Harter Fell

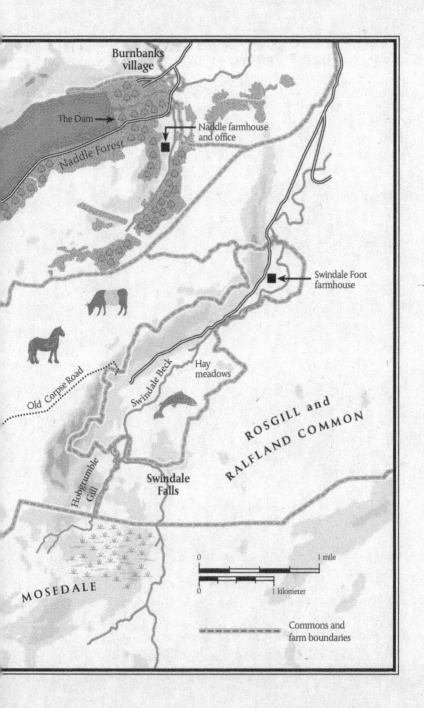

PLACE NAME GLOSSARY

BEASTMAN'S CRAG (near Swindale): *The rocky height of the cattle-man* (Old English/Old Norse)

BECK: *Stream* (from Old Norse; *bekkr*)

BENTY HOWE (Bampton Common): *The hill where the bent-grass grows* (Old English/Old Norse)

BLÅFJELLENDEN (Fidjadalen, SW Norway): *The end of the blue mountain* (Norwegian)

BLEA WATER (Mardale Common): *The dark lake* (Old Norse/Modern English)

BRANT STREET (Mardale Common): *The steep path* (Old Norse/Old English)

BURN: *Large stream or small river* (Scotland)

CARRIFRAN GANS (Scottish Borders): *The fort of ravens* (Celtic)

CATSTYCAM (above Ullswater): *The steep path frequented by wild cats* (Middle English/Old Norse)

COCKLAKES (near Matterdale): *The place where the black cock play* (Old Norse)

COCKLE HILL (Bampton Common): *The hill where the black cock play* (Old Norse)

COLEDALE: *The valley of the charcoal burners* (Old Norse)

COMMON: *Unenclosed pasture for communal use* (Middle English)

CORRIE: *An amphitheatre-like valley formed by glacial erosion* (from Gaelic; *coire*)

CRAG: *Rocky height, major outcrop of wall or rock* (from Gaelic; *creag*)

DALE: *Valley* (from Old Norse; *dalr*)

FELL: *Hill, mountain, tract of high unenclosed land, high ground* (from Old Norse; *fell*, meaning a single hill and *fjall*, meaning mountainous country)

FJELL: *Mountain* (Norwegian)

FORCE: *Waterfall* (from Old Norse; *fors*)

GILL/GHYLL: *Ravine with stream* (from Old Norse; *gil*)

GLEDE HOWE (near Swindale): *Kite Hill* (Old Norse)

GOUTHER CRAG (Swindale): *The rocky height with an echo* (Old Norse)

GRAN PARADISO: *Great Paradise* (Italian)

HARE SHAW (Mardale Common): *Hare wood or copse* (Old Norse). This hill is no longer wooded.

HARTER FELL: *The mountain frequented by deer* (Old Norse)

HAWESWATER: *Hafr's Lake*. Hafr being a personal name (Old Norse)

HERON CRAG (Mardale Common, and many other Erne, Iron, Heron and Aaron crags elsewhere in the Lake District): *The rocky height frequented by white-tailed eagles* (from Old English; *erne*)

HIGH STREET: *The high paved road* (anglicized from the original Latin name, Via Alta)

HOBGRUMBLE GILL (Swindale): *The rumbling ravine haunted by a hobgoblin or bogle* (Old Norse)

KIDSTY PIKE (Bampton Common): *The peak at the top of the steep path where the young goats go* (Old Norse)

MARDALE: *The valley of the lake* (Old Norse)

MATTERDALE: *The valley where the bedstraw grows* (Old Norse)

MELL FELL (Matterdale): *Bare hill* (Cumbric)

MOSEDALE (above Swindale): *The valley with a bog* (Old Norse)

NADDLE: *The wedge-shaped valley* (Old Norse)

RIGGINDALE (Mardale and Bampton Common): *The valley below the ridge* (Old Norse)

RIVER LIZA (Ennerdale): *The bright, light river* (Old Norse)

ROSGILL (near Swindale): *The ravine where the horses graze* (Old Norse)

ROWANTREETHWAITE: *The clearing with the rowan tree* (Old Norse/Modern English)

SWINDALE: *The valley where the pigs graze* (Old Norse)

TARN: *Small mountain pool* (from Old Norse; *tjörn*)

ULTHWAITE RIGG (near Swindale): *The ridge above the wolf clearing* (Old Norse). There are many places in the Lake District that include *úlfr*, the Old Norse for wolf

WOOF CRAG (Mardale Common): *The rocky heights frequented by wolves* (Old English)

LIST OF ILLUSTRATIONS

INTRODUCTION

The Willows

HARTER FELL:
The mountain frequented by deer
(Old Norse)

Three straggly trees huddle on the fissured face of Harter Fell, high in the Lake District's eastern corner. From thirty feet below, I can tell that they are willows, but I need to get closer to work out which species. With plenty of well-rooted birch and rowan to cling to, scrambling up the first part of the flower-hung crag is easy enough. The upper section, a few more degrees to the vertical, is a different matter. A loss of footing now would result in a slide, a plummet and a limb-breaking landing on the boulder scree below.

Few people set foot up here, leaving botanical treasures undiscovered in steep gullies and on brittle ledges. Water running over the crumbly, calcium-rich rocks forms patches of fertile rudimentary soil; great for flowers, not so great for climbing. Halfway up this slippery mess of a cliff, a solid-looking foot-long triangle of rock comes off in my hand, shattering on the ground below. I retreat with my pulse racing.

Working around the base of the crag, I find a gentler ascent via a narrowing grassy corridor, hemmed in by rising cliffs. As usual, my progress is slow, distracted every few steps by the summer wildflowers. Stone bramble, alpine lady's-mantle, northern bedstraw, lesser meadow-rue and starry saxifrage enliven the rocks, the poetry in their names adding to the allure of their shapes and colours.

The willows come back into view with just a narrow sloping slab bridging the gap between us, which I inch across, crab-fashion. I feel as if I'm reaching hallowed ground. Growing beneath the unruly mesh of the willows' branches are the fleshy stems of roseroot, the smooth oval leaves of devil's-bit scabious, sturdy heather and vivid bilberry, a fragrant feast for any herbivore. That these plants are still here tells me that none have been brave or hungry enough to make the crossing. The soil is light and fluffy, with the rich mossy smell of an ancient oakwood. Any boot marks I leave will be quickly colonized, the plants erasing the evidence of my visit.

Their glossy green foliage tells me that these are tea-leaved willow, a species which grows in only a handful of places in the Lake District fells. No more than shoulder-height, their squat

Tea Leaf Willow

and sprawling form keeps them stable on their wind-battered refuge. The promise of such rare and beautiful plants makes botanizing in these crags intensely addictive. There's always one more ledge to investigate, one more unexplored gully with secrets to uncover. I've never made it back to the car anything less than hopelessly late.

Fast growing and producing huge quantities of tiny airborne seeds, willows are pioneers, quickly establishing in damp ground where the conditions are right. For thousands of years, willow bark has relieved human suffering – its soothing power led to the development of aspirin. Perhaps herbivores can detect this medicinal benefit, adding to willow's edible appeal. Smaller creatures are just as enticed; willows are second only to oaks in the number of insect species they support.

What I love most about willows is their regenerative ability. Stick a cut branch into damp soil during the winter months, and by spring, it will have sprouted leaves, on its way to becoming a whole new tree. Even a small cut branch kept inside on a tabletop can develop bright buds in spring, tiny green symbols of hope and renewal. There is no tree as hungry for life as a willow.

Like many other palatable species in this open landscape, these tea-leaved willows have become relics, fortunate to have found a place that gives security through inaccessibility. But their security doubles as a prison. In their isolated eyrie, their seeds can only blow into the grazed land below, where they have little hope of growing. I will return in winter, when the sap isn't flowing, to take cuttings and plant them somewhere they stand a better chance.

As happens so swiftly in the mountains, the weather is turning. Grey skies are bleeding into blue and a fine drizzle is making the slab even more precarious. Before daring the reverse scramble, I sit to absorb the view, wind swirling the rain into my face.

Before me is the four-mile crescent of Haweswater Reservoir, lying in the palm of Mardale, a name given by Vikings that translates as 'the valley of the lake'. The word *dale* comes from the Old Norse word for valley: *dalr*. It's one of many local geographic terms with Scandinavian roots, indelible reminders of this land's long and ever-changing human history.

On a clear day, I'd be able to see further, across the green farmland of the Eden Valley to the Pennines, the long chain of hills that makes up the eastern horizon. Today the veil of rain that has drawn across Haweswater's far end limits my view to the land into which I've been sinking roots for the best part of a decade. Sheltered from the wind by the crag wall beside me, I look out over a landscape littered with memories: the dinosaur-hunting expedition into the steep wooded gorge of Rowantreethwaite with my seven-year-old son; cooling swims below Swindale Falls following long days tramping the bog in Mosedale; carrying an injured lamb on my shoulders after gathering the flock from Riggindale; the new patch of bird's-eye primrose I'd found in a flush on Mardale Common. Each experience a new intimacy.

Mardale is a valley where every lichen-encrusted rock and moss-clad tree resonates with stories. The remains of a flooded village lie beneath the reservoir's dark surface. The first golden eagles to nest in England in a century did so on the crag where I'm perched. This is a place of ruins and wildness, of traditions ancient and modern. It is a working landscape that's not working as well as it could be. It's also a landscape of hope, where change is afoot. As the site manager for RSPB Haweswater, I'm one of many tending to the land as best I can to nurture its wildlife, to help it function better and undo some of the damage that we humans have inflicted on it over the centuries.

The hulking mass of Harter Fell, my mountain lookout,

brings the valley's only road to a dead end. Running from here, between drystone walls, it follows the curve of Haweswater's eastern shore as far as the old dam workers' village of Burnbanks, five miles north at the valley mouth. Working at this far southern end meant I revelled this morning in driving the reservoir's full length. The commute never gets old.

From where I sit, I can just pick out the ancient woodland of Naddle Forest clothing the eastern slopes above the water. In spring, the delicate flowers of wood anemones, bluebells, wood sorrel and yellow pimpernel stitch the forest floor. Above, the branches of oak and ash, downy birch and hazel, clothed in moss and ferns, are highways for the local red squirrels. The terrain there is so steep and the plant life so riotous that the forest receives few human visitors. I've spent happy days exploring those woods, navigating steep ground in search of parasitic bird's-nest orchids or tree lungwort lichen, listening out for wood warblers or redstarts, each new species record contributing a little more to the detailed picture of this place and its wild inhabitants.

Even though the M6 motorway is only seven miles away, Naddle Forest can feel like an escape into wilderness, a total immersion in mossy chaos. Yet this forest is a relic too. Big enough to lose yourself in, perhaps, but a fragment of a much larger woodland that historically extended further up the now treeless hills and down below the waterline.

Wallow Crag erupts from Naddle Forest's canopy, and from its top one of the valley's starkest contrasts is revealed. For while the eastern side is rich with woodland, the fells that rise to the west, half a mile across Haweswater's wind-chopped water, are bald but for a scatter of conifer blocks. The open ground is stripped, its rocky skeleton breaking through a skin of yellow, green and brown.

My Harter Fell willows are the tattered survivors of a habitat that has effectively vanished from the Lake District. Before the epoch of intensive grazing, montane scrub formed extensive patches of willows and wildflowers just below the windswept fell tops. I've spent time in montane scrub in Norway where it covers hundreds of square miles. With an understorey of many of the same bulky and nectar-rich flowers that grow on the crags of Harter Fell, the habitat supports a spectacular array of wildlife, from bluethroats and ring ouzels to black grouse and golden eagles. Our hills, by comparison, are just a great blank space, haunted by the shadows of vanished creatures.

This is the landscape my colleagues and I have been charged with looking after, a landscape of shattered fragments, longing to be put back together. Aldo Leopold, one of the godfathers of modern nature conservation, said, 'To keep every cog and wheel is the first precaution of intelligent tinkering.' The Harter Fell willows are one such cog.

PART I

Imperfection

There was a time when meadow, grove, and stream,
  The earth, and every common sight,
      To me did seem
      Apparelled in celestial light,
    The glory and the freshness of a dream.
It is not now as it hath been of yore;–
      Turn wheresoe'er I may,
      By night or day,
The things which I have seen I now can see no more.

William Wordsworth, from
'Ode: Intimations of Immortality from
Recollections of Early Childhood', 1804

CHAPTER I

The Eagle Hut

RIGGINDALE:
The valley below the ridge
(Old Norse)

Tonight, I sleep with the ghosts of eagles. It's late April and I'm due at the head of Riggindale before sunrise tomorrow to find out if ring ouzels are nesting in the corrie of Sale Pot this year. Until then, the Eagle Hut, a small wooden shed nestled in the crook of a lichen-spattered stone wall, is my home for the night. It's not much to look at. Held down by steel cables, the Eagle Hut's single-pitch roof and scarred timber sides bear witness to decades of pummelling by an upland climate. Its setting, though, is a genuine feast for the eyes.

Riggindale is like a diagram in a geology textbook, its U-shaped glacial trough running perfectly west to east and its steep back wall topped by a pencil-sharp skyline. It is a valley of two halves, with sheer cliffs and steep ravines on its southern side punctuated by twisted birch and rowan trees. The north side is bald rock, grass and bracken, its gentler gradient giving easy access to sheep and deer. The Eagle Hut sits near the valley's mouth, beside Riggindale Beck which snakes its way down towards Haweswater's western shore in extravagant, boulder-strewn curves.

I scramble down behind the hut to reach a small pool in the beck to fill a bottle to cook my dinner. Crossing a flush, where water seeps over the ground, I notice red-tentacled sundews and the lime-green stars of butterwort amid the triangular stems of sedges and a lush moss carpet. Where the flush meets the beck, I find rosettes of the heart-shaped leaves of grass-of-Parnassus. In the coming weeks, long graceful stems will rise from these rosettes, bearing clusters of pure white flowers, delicate tracery on each of their five perfect petals. This is a plant that belongs here, so much so that it has been adopted as Cumbria's county flower.

I can't claim such a long connection. I was born in Scotland, but raised in deepest Devon, in a mud and straw house folded into the landscape by gently rolling hills and towering hedges. My brother and I enjoyed a free-range childhood that feels more like make-believe the further I get from it. We ran wild in the self-contained world between the two ends of our winding lane.

Our parents worked hard running a small business that

Grass of Parnassus

made and distributed West Country produce to London, which always provided work for the two of us. When we weren't on scotch-egg duties (two pence per peeled egg) the primrose and fern-filled hedges, fields and lanes were our domain, supplying blackberries in the summer, hazelnuts in the autumn and hiding places all year round.

Spending my formative years surrounded by nature meant that I never paid it much attention; it was just there. I assumed that everyone had the chance to pick tiny wild strawberries as they walked home from the school bus, that everyone made dens in thickets and caught grass snakes and slow worms. It was only moving away that made me realize how lucky I'd been.

⟡

As I walk back over to the Eagle Hut in the fading light, my shadow lengthening, I remind myself how fortunate I am to be able to call this work. It is not a privilege I wear lightly, and every day I'm striving to be worthy of this place and the stewardship role I've been entrusted with.

The sun is sinking over High Street, the long ridge at the far western end of the valley. Its shadow comes galloping towards me, stripping away the day's warmth as it passes. I need to eat and sleep. Tomorrow starts early. As I unpack my cooking things, I find a faded RSPB membership leaflet under the bench that will be my bed for the night, harking back to the days when this shed was more than just a dusty place to sleep. Over the years, thousands of people made pilgrimages to this valley, for it was here, among Riggindale's remnant woodland, screes, crags and flushes, that England's only golden eagles made their home for nearly five decades.

Eradicated in England by the Victorians, golden eagles

began to show up again in the Lake District soon after it became a national park in 1951. A handful of birds made nesting attempts in remote locations, but throughout the late 1950s and early 1960s their nests were left unfinished and eggs weren't laid. It turned out that sheep dips, used to protect against insect pests, were scuppering the eagles' breeding success. Sheep would often die whilst grazing out in the fells, and their carcasses made up the bulk of the Haweswater eagles' diet. And so the noxious organochlorine chemicals in the dips made their way into the eagles' bodies, weakening would-be mothers and thinning the shells of their eggs. When organochlorine dips were finally banned, the birds recovered, and in 1969 a pair laid the first golden eagle eggs England had seen in 170 years on the steep north face of Harter Fell, a stone's throw from where I had found the tea-leaved willows.

Things didn't go well that first year. The Mardale Head car park, just below Harter Fell and its pioneering nest, generated more disturbance than the eagles could bear and two precious eggs were abandoned. Riggindale, a mile around Haweswater's south-west shore, proved more tranquil, and it was here that the birds were to have more success, raising sixteen chicks between 1970 and 1996. Not only was Riggindale quieter but it also happened to have an 'Eagle Crag' near the valley mouth, just a few hundred metres from the hut, named centuries ago to mark a historic nest site. The eagles' instincts had drawn them to a place that had been used by their kind since time immemorial.

The RSPB arrived soon after the eagles settled. Establishing an office in one of the wooden houses at Burnbanks at the northern end of Haweswater, my predecessors worked tirelessly during the decades the eagles spent in Riggindale to ensure that the birds had the best possible chance of rearing their young in peace. For a short time, there was even a second nest, over in the western Lake District. Without the protection afforded by the RSPB, this territory didn't fare so well and was soon abandoned, leaving the Riggindale eagles the gloomy accolade of being the last of their species in England.

The Eagle Hut's initial use was as a base from which to monitor the nest and guard against thieves intent on stealing the birds' eggs, a real danger in those days. Later, it became the Eagle Viewpoint, where a committed band of wardens and volunteers armed with telescopes helped visitors to see these living embodiments of wildness in their only English outpost.

Golden eagles are long-lived birds, and just three overlapping generations made up the lonely lineage at Haweswater. After the male of the original pair died in 1976, he was replaced by a second male who went on to become Britain's oldest recorded golden eagle, living for thirty-two years. The original female passed away in 1981 and was replaced by another newcomer, who took up with the second male. The most tragic member of this little group was the third and final male, who arrived in the early 2000s. He spent the first couple of years paired with the second female, but she was now so elderly that she failed to produce any eggs. After she died in 2004, he was left in the valley by himself. Every spring he would perform his tumbling, undulating display flight, in the hope of attracting a new mate, but by this time the closest golden eagle population in Southern Scotland had dwindled, and there were simply not enough birds in the vicinity. From time to time, rumours of other golden eagles in the Lake District would circulate, but England's last golden eagle never succeeded in pulling in a mate. With every year that passed, his breeding display flights seemed less and less enthusiastic, and eventually in 2015 he vanished, having spent over a decade alone; a potent symbol of wildness bleeding out of the landscape.

I started at Haweswater in 2013, just before this nadir. For my first couple of years, we continued to operate the Eagle Viewpoint during weekends in the spring and summer. It was always an incredible thrill to catch sight of our golden eagle, even if distantly. There is no more spectacular bird. Even though they aren't now the UK's largest bird of prey, having been beaten to the top spot by reintroduced white-tailed eagles, their sheer power and grace makes them incomparable. With two-metre wingspans, and weighing more than a newborn human, they can kill prey as large as young red deer – their tendency to take lambs and game birds is why they've had such a history of persecution. Yet for all the excitement of seeing our eagle, working at the Viewpoint was always tinged with sadness; we knew what was coming.

He was about twenty years old when he disappeared, a typical lifespan for wild eagles. That winter there had been terrible storms, creating intolerable conditions for a bird already past his prime. Despite his keen eyesight, he would have struggled to pick out prey through the endless rain, the wind constantly blowing him off target. He most likely died of starvation, tumbling from his craggy perch to be ingloriously pulled apart by foxes and ravens.

The same angry weather in the winter of his demise ripped off the Eagle Hut's roof. The storms unleashed floods, forcing hundreds of people from their homes across the county. Stefan, a Polish war hero who settled in the area and dedicated over twenty years of his life to protecting the eagles in the 1980s and 1990s, passed away in his sleep a few days after the last eagle sighting. Something more than the death of a bird had occurred; a vital thread had unravelled.

That the whole of England is incapable of supporting a single pair of golden eagles, where historically there had been hundreds, should be a source of national shame. I felt some of this shame personally – after all, England's last eagle had disappeared on my watch. Although the rational part of my brain knew it was inevitable, this knowledge came at a time when my job was already taking a heavy toll.

We patched the Eagle Hut up, put its roof back on, perhaps unable to accept that the reign of the Riggindale eagles was over, or as a mark of our belief that they would return. It may just be a wooden shed, but as I settle down to sleep in its creaking confines, the Eagle Hut's history and significance are palpable, as is Riggindale's sense of vacancy. Buzzards and ravens, birds that had been chased out by the eagles, soon moved back into the valley. Wonderful though these species are, they are poor replacements for their absent king.

In the night, I'm woken from a dream of wind-ruffled feathers by a group of red deer. The wind has picked up and they have come down the valley from the higher fells above to seek shelter behind the walls. I drift back off to the sound of their breathing and shuffling outside. My alarm goes off at 4 a.m. There is no sign of the deer by the time I emerge. I head west, following the beck upstream as it gurgles in the darkness.

☙

Ring ouzels are the mountain-dwelling, white-collared relatives of blackbirds. Most spend the winter months in the Atlas Mountains of Morocco, returning to upland areas in Northern Europe to breed. Like a depressingly long list of other birds, ring ouzels are on the red list of species of conservation concern, their numbers having declined sharply over recent decades. Living in such steep and remote terrain, they are more often heard than seen, and with their tendency to stop singing at sunrise, finding them requires an early start. Sleeping in the Eagle Hut has shortened my nocturnal hike.

The location for my survey this morning is a small hanging valley, or corrie, called Sale Pot, at Riggindale's north-western corner. Riggindale Beck flows out of Sale Pot, so keeping it to my left and working upstream makes the corrie easy enough to find, even in the pre-dawn darkness. It's a gentle climb for the first mile or so, but as I reach the short, steeper section up over the corrie's lip, the sky starts to lighten so I pick up the pace.

It's curious, Sale Pot. As I enter it, I always expect to be greeted by a tarn, a small mountain lake, like the ones which occupy two similar corries just over the ridge to the south. There's no doubt that there used to be water here; Sale Pot's name means 'willow pool'. But there's no tarn now, just a tell-tale network of drains dug by people long dead to gain access to deposits of peat which they could cut for fuel. Other than a single, stunted bush growing up on a high ledge, there aren't any willows here either.

No matter how much Sale Pot might have changed, the ring ouzels seem to like it. As soon as I enter its glacially carved amphitheatre, I can hear the distinctive double whistle of the male ouzel echoing around its steep sides. I take a seat on a flat rock, hoping that the gathering light will allow me a glimpse of the singer. As I scan with my binoculars, there's movement on the opposite side, above where the corrie's tarn must once have been. I see a badger lumber along a sheep trod and dis-appear into its sett in the boulder scree. I'm 550 metres above sea level, in a virtually treeless landscape. Not where I'd have expected to see a badger. Nature is full of surprises.

The ring ouzel keeps singing until the sun comes up over the ridge behind me, and right on schedule he shuts up. Shortly after, both he and his mate appear, hunting for worms and grubs in the short grass at the base of the corrie wall. They are seriously smart-looking birds. Against his dark feathers, the male's crisp white markings are like an Incan necklace. The female is marked in the same way, but muted, as if she's spent too much time in the mountain rain.

Although not a mainstay, ring ouzels are one of many species that would have fallen prey to Riggindale's eagles. In more pristine upland landscapes red and black grouse, their relative the ptarmigan and mountain hares make up the bulk of a golden eagle's diet. Of those, only red grouse can be found in the Lakes, and even they are few and far between. But golden eagles are a highly adaptable species and will scavenge if wild prey isn't readily available. At Haweswater, the eagles survived predominantly on the carrion of both sheep and deer. They'd also hunt rabbits, red grouse and the occasional crow, ring ouzel, badger, fox or any other creature unwise enough to let its guard down.

The rise and fall of the Riggindale eagles feeds Haweswater's wild and rugged aura. It isn't wild, though. Perhaps more so than anywhere else in the Lake District, this is a landscape that has been dramatically moulded by human hands.

Leaving the ouzels, I follow the beck back down towards the hut to pack up. Other than the remains of an old stone barn, the hut is the only building I can see. A century ago, there would have been a church with a sturdy square tower in my line of sight, and cottages and several small clusters of farm buildings surrounded by fertile farmland.

For centuries, Mardale and its two small hamlets, Mardale

Green and Measand, had been on an important trade route connecting the town of Penrith to the north with Kendal and Ambleside to the south, thanks to two mountain passes over which livestock and goods flowed. Mardale Green's church, parsonage and the small but comfortable Dun Bull Inn spoke of the area's prosperity, but as trains and cars replaced feet and ponies the valley became a tranquil backwater. Yet as Mardale's community continued the quiet pastoral way of life that had sustained them for generations, distant powers were making designs on their valley.

Some 75 miles south of Mardale as the eagle flies, the burgeoning urban population of Manchester and its booming industries had a thirst for water that couldn't be satisfied by local rainfall. So, in 1919 an Act of Parliament was passed that enabled the Manchester Water Corporation to acquire 10,000 hectares of land in and around Mardale, to construct a dam and create a reservoir.

Before the dam, the original Haweswater Lake was smaller, shallower and partially divided by a spit of land into two connected water bodies, which the old maps name as High Water and Low Water. Haweswater Beck flowed out of Low Water at its eastern end, where stepping stones allowed people to cross its great width. By 1935 the 470-metre-wide, 36-metre-high concrete bulk of the dam across the valley was complete, the flow in the beck was arrested and the water started to rise. The valley floor, its farms, fields and hamlets were slowly lost to the rising water, which today sits 29 metres above its original level. The new reservoir kept the name Haweswater, given to the lake by people who could never have imagined its modern incarnation.

The desecration of Mardale and the displacement of its forty residents generated huge controversy and opposition; it was the HS2 of its day. In the end, uprooting a handful of families was considered a price worth paying and no amount of public outcry, no matter how strident, could stop the march of progress. Alfred Wainwright, the Lake District's most famous fell walker and popularizer, first visited Mardale in 1930, just as the construction of the dam was getting underway. He never got over the valley's modification, referring to the 'rape of Mardale' in his final book.

However, alongside the human cost came great reward; Manchester's thirst had been sated. The engineering that conveys the purity of Lakeland rain all that way to Mancunian faucets transformed the lives of vast numbers of people, providing reliable access to clean drinking water that hadn't been available previously. Haweswater today remains the most important reservoir in the North West of England, supplying two million people with their daily drinking water – but that wouldn't have provided much consolation to the folk whose houses were now underwater.

So complete was the alteration of Mardale that even its name has been partially lost, with most people, myself included, now referring to the valley and the area as a whole simply as Haweswater. Its history lends the area a melancholy air, particularly when the footprint of flooded settlements, roads and bridges becomes visible during exceptionally dry summers. The drystone walls at the reservoir's edge, boundaries to drowned fields, are a year-round testament to the old landscape. They follow the slope, lost to sight as they plunge into the dark water. Stones that had once kept sheep in place now provide refuge for fish.

In most Lakeland dales the valley bottom with its lake or river grades gently through sloping woods and pastures to the mountainous land above. The raising of the water in Mardale and the resulting transition straight from lake to fell is what gives the valley its wild, elemental feel. In part, the clearing of the valley's people made room for the golden eagles, which shun humans more than any other bird. They wouldn't have put up with the bustle of Mardale's now flooded settlements. Like the villagers, they too have now been consigned to history; this is a valley with more than its fair share of ghosts.

Although the eagles are gone, the RSPB have remained at Haweswater. United Utilities is the latest incarnation of the Manchester Water Corporation, North West England's water company, and owner of the reservoir and its 100-square-kilometre catchment. The partnership between our two organizations has been evolving ever since the RSPB arrived to watch over the eagles. In 2012 that relationship took a big step forward when we took over the tenancies of Naddle and Swindale, two hill farms that were about to be vacated by long-term farming tenants. Soon after, I was employed to oversee their running.

Naddle and Swindale Farms occupy neighbouring valleys on the eastern side of the reservoir. As well as 750 hectares of enclosed land, complete with sheep flocks, barns and farmhouses, our tenancies give us grazing rights on three upland commons, large, unfenced expanses of fell land. Mardale Common includes all the land between Naddle and Swindale and the mountainous terrain that overlooks the reservoir's southern end, including Harter Fell and half of Riggindale. Over on the western side of the reservoir, Bampton Common's 28 square kilometres of rough, high ground extend north from Riggindale Beck. Rosgill and Ralfland Common is lower lying, its boggy acres running east from Swindale towards the road-straggled village of Shap. These tenancies and grazing rights give us a major stake in the management of over 3,000 hectares of land containing a mix of bog, heath, grassland and scrub, watercourses, tarns, crags and moor.

What makes working at Haweswater unique is that we are aiming to find a solution to one of the great environmental conundrums: how to rebalance farming and nature. Over 70 per cent of the UK's land area is used for agriculture – a higher proportion than almost any other country – and nature finds herself increasingly restricted to the spaces between farms. So whilst conservation projects and nature reserves have a vital role to play, the fact is that their efforts will be in vain if we can't find a way for nature to thrive on farmland too.

What we're trying to do, and the reason for taking on the farm tenancies, is to develop a way to look after our land that occupies the middle ground between hill farming and conservation, restoring nature, respecting traditions, producing food, and supporting the rural economy. By doing that, and then showing how we are doing so to others, we have an opportunity to effect huge change.

From the start, I knew I had one of the best jobs on the planet, one in which no two days are the same. The most rewarding are those when I get my hands dirty hanging a gate, digging a pond, or planting trees and wildflowers. I get to stretch my legs checking on our sheep, cattle and ponies, or the state of the miles of fences and walls that are supposed to keep them in place. The spring and summer months often find me in the fells counting birds, butterflies and plants, while in the autumn I head out in search of seeds for our nursery.

It's a collective effort, working shoulder to shoulder with a team of people who are every bit as driven as I am. Spike, Bill and Dave had already spent decades at Haweswater before I came into the job, and sometimes I still feel like I'm playing catch-up. New colleagues have swelled the ranks; Heather, Jo, Ashley, Richard and Bea have brought fresh ideas and enthusiasm, while Matthew and Murray and their team of farm dogs exercise their skill to keep the livestock fit and healthy. A friendly mob of volunteers, some of whom become residents with us for weeks or months at a stretch, give generously of their time and energy to lend nature a helping hand. Together, we are constantly learning, trialling and tweaking in order to steward our mosaic of habitats towards greater richness. It's an amazing feeling to stand in the middle of our land, knowing our efforts will influence nature from skyline to skyline.

But it's not all bracing walks in spectacular scenery. I spend a lot of my time in the centuries-old farmhouse at Naddle Farm, into which we've shoehorned our office and volunteer accommodation. Naddle can be a bustling place; wellies in the porch, several rows deep; the radio blaring as the sheep are shorn; swallows and house martins swirling overhead. It's here, hunkered over the computer, warmed by a log burner, that much of my daily grind takes place. After the morning tea and chatter subsides as the rest of the team head out to put in the physical graft, I have reports to write, budgets to keep in line, grants to apply for and all the bureaucracy that comes with working for a large NGO.

On occasion, it's a job that's been downright unpleasant. The Lake District is a place of time-honoured traditions and of small family farms. Unsurprisingly perhaps, the prospect of an institution like the RSPB taking on a tenancy, let alone two, was so far from the norm as to be considered close to criminal in some quarters. When the RSPB did so in 2012, my colleagues and I effectively became farmers, but that didn't mean we were automatically welcomed by everyone. By some in the community we were perceived as cogs in the machinery of a faceless organization, imposing external ideas that didn't sit well with the area's history or way of life. I understood that.

At times, however, the discomfort with our presence in the valley spilled over into open hostility. Like most farms, we rely on casual labour to help at lambing time, with gathering sheep and at other key points in the farming year. Some of the local shepherds we employed demanded unreasonable pay hikes and threatened to walk away if we didn't oblige. At open days and walks that we organized, we were regularly told that we were idiots and that we didn't know what we were doing. Rumours and gossip swirled around us, mostly fuelled by the suspicion that we were about to remove all the sheep. For some in the Lake District there could be no greater sacrilege.

Becoming part of this sprawling, complicated, beautiful, troubled landscape is an honour. I have played a part in shaping its hay meadows, woods, bogs and streams. I know where the mountain ringlet butterflies breed, where the last patch of alpine saw-wort grows, where the best places are to collect bitter vetch seedpods and where the peregrines nest. I also know how much more work there is to do and how much more widespread the pockets of good habitat need to be before we'll have a landscape that's fit for eagles again.

Comprehending this challenge involves understanding Haweswater's ecological state along with its past, present and possible future. I've had to learn about the intricacies, the small and fragile lives of plants and animals, and how they fit into the bigger picture. Developing this understanding, shaping a plan and then enacting it, have become my life's endeavour.

CHAPTER 2

Lonely Flowers

BLEA WATER:
The dark lake
(Old Norse)

It is a wide blue May morning, and I am setting out to visit England's rarest mountain flower. Pyramidal bugle, *Ajuga pyramidalis*, teeters at the edge of national extinction in a location known only to a few individuals – I'm not going to tell you exactly where.

I'd been asked to check on the *Ajuga* by master Cumbrian botanist Jeremy Roberts, the unofficial guardian of this vulnerable treasure. The call came after several years of getting to know Jeremy and convincing him that I was a committed and sufficiently knowledgeable naturalist to be admitted into his circle of trust. I'd asked him for directions to find the *Ajuga* a couple of years earlier, but I obviously wasn't ready, and had been courteously fobbed off. At the time, I couldn't understand why on earth he was being so cagey. I knew that the *Ajuga* was incredibly rare, but his protectiveness seemed disproportionate. Surely no one steals rare plants any more.

Having developed my knowledge of Lakeland botany over

the course of years, I finally earned my stripes along with Jeremy's trust. Thanks to his holiday plans clashing with the plant's usual peak flowering period, I was given the closely guarded directions to *Ajuga*'s last redoubt, along with a strict directive not to share them with anyone and instructions of what to record in order to carry out the fragile colony's annual health check.

This morning my walk to the little plant's home passes through classic Lakeland scenery, all soaring peaks and rocky streams, shared only with the skylarks, meadow pipits and sheep. Following Jeremy's written directions is a treasure hunt through unfamiliar territory. Instead of a map, I must follow descriptions of subtle features in the landscape. Navigating the corries and coves, mine workings and waterfalls, I could be heading for Smaug's cave.

The Lake District is England's largest and most visited national park. At 2,362 square kilometres it covers around a third of Cumbria, England's most north-western county. The national park contains the country's longest and deepest lakes, its three highest mountains and hundreds of other lesser peaks. Mountain denizens elsewhere in the world scoff that our Lakeland hills are a little on the small side compared with the Alps, the Rockies or the Himalayas. But if the Ordnance Survey calls them mountains, who are we to argue?

Most of the Lake District's millions of annual visitors come for the grand views, but landscapes like these operate at multiple scales. When I walk in the hills, or anywhere else for that matter, my attention is split between the macro and micro, between the panorama and its texture, the scene and its characters. The countryside contains so much more than you see on a postcard. It is an intricate composite of tiny parts, stitched together into a living tapestry.

It's a weekday and I set off at first light, so I barely see another person on my walk out. With my eyes to the horizon, my spirits soar with the feeling of having the fells to myself and the promise of discoveries ahead, but there is an altogether different sensation when I look to my feet. For step after step after step, the ground is almost completely devoid of flowers. During one section of the walk along a ridge, I make a point of counting. In an hour I record just one open flower, a solitary marsh marigold on the bank of a burbling ghyll. The only other plants covering the ground are coarse grasses and rushes, dominated by just three species: moor mat-grass, heath rush and deergrass. This botanical bleakness is par for the course for much of the Lake District.

As I approach the crag that holds the fragile *Ajuga* colony, the plant life picks up. Where the ground is more broken, different species start to appear. The silvery-edged leaves of alpine lady's-mantle cling onto rocks and yellow mountain saxifrage picks out the stream edges. In more inaccessible places, bitter vetch and sea campion dangle over ledges, perfectly poised to cast their seed down onto the ground below later in the year. Heather and bilberry grow richly on vertical terrain, along with ferns, great wood-rush and occasional stunted birch and rowan trees.

After some scrambling, slipping up and down scree slopes and peering into rock clefts, I find what I have come for. Small and unassuming, pyramidal bugle is a flower for connoisseurs. It's a member of the mint family and recognizably a close relative of the more familiar bugle species, *Ajuga reptans*, a plant common in many habitats including gardens. Pyramidal bugle has a lot more class. For most of the year it survives as a low-growing, dark-green rosette, which gives rise each spring to a pyramid-shaped flowering spike on which tiers of small

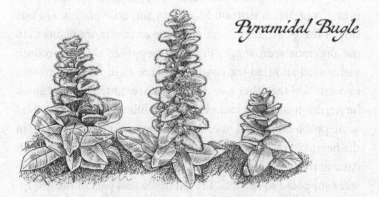

Pyramidal Bugle

purple flowers lurk beneath hairy leaves, which become more purple themselves as they near the pyramid's peak. The covering of fine hairs makes the whole plant appear to be in film-star soft focus.

At this, *Ajuga pyramidalis'* solitary English station, the plant grows on a series of narrow ledges, either side of a steep, scree-lined gully. James Backhouse junior, one of the Victorian giants of natural history, and a correspondent of Charles Darwin, discovered *Ajuga* here in 1869. His shadow looms long in the history of mountain botany, and he first described many of the Lake District's alpine plant localities.

For reasons unknown, the colony's precise location was then forgotten, and by the time Jeremy rediscovered it in 2006, his was only the fourth visit in 137 years. Locations for ultra-rare species like this one are often kept secret by local botanists, with details given to only a handful of trusted initiates. With improvements in the accuracy of recording geographic information and the ease of sharing it, many more people have visited this spot in recent years. That of course has its pros and cons.

My job today is to count the *Ajuga* plants, record the condition of vegetation surrounding them and take photos. On one side of the gully, there seem to be one or two more plants than the previous year, suggesting that they were able to produce viable seed and that the conditions were right for new growth to occur. On the other side, the plants are dotted around and a bit harder to assess. It looks like the nibbling of voles has caused some problems, and in one area the shade cast by an increase in the height of the surrounding bilberry looks like it might be an issue in future. However, I find plants in most of the places they were supposed to be, until I reach the lowest part of the group.

I've brought photos from the previous colony survey as a reference point. Holding up the relevant picture against the lowest ledge, I am momentarily confused. Where last year there was a healthy, vigorous plant, there is now only a trowel-sized depression in bare soil. I stand back to check that I'm in the right place, taking in the steep terrain, inaccessible to any grazing animal. I double-check the photos, look at the crag from different angles and try to come up with alternative explanations for what I am seeing. It is no use. That gap in the soil means only one thing; the plant has been stolen, complete with roots and soil.

The chances of someone stumbling upon pyramidal bugle here are incredibly slim. It's growing on near-vertical, rocky ground, well away from any footpath and a long steep walk from the nearest parking spot. Though attractive, it's hardly showy, not exactly a species that would make much of an impression in a garden. Whoever stole this plant must have been a collector of rarities, who knew where it was and had set out specifically to rob it. Jeremy's insistence on secrecy was justified.

Pyramidal bugle should not be in this vulnerable state. It is not a species with particularly exacting requirements. It grows widely in mountain regions in Europe, in heathland, grassland and on rock ledges. I've seen it growing in south-western Norway in woodland clearings and old hay meadows. Across a swathe of north-west Scotland, it can be found growing right down to sea level.

Because of the location of this tiny Lake District colony at the very base of a crag, like the willows on Harter Fell, any seeds that fall from the parent *Ajuga* plants can only drop into areas that are grazed. Seedlings are seen in some years, but every time without fail these hopefuls are eaten. My friend Simon – a fellow mountain-plant obsessive – has a stock of young *Ajuga* plants grown from seeds collected from the crag a few years ago, in readiness for future efforts to help boost the plant in the wild.

Its habit of not flowering every year means that *Ajuga* doesn't have the potential to become a prevalent wildflower like its common bugle cousin, but there is nothing about its biology that should have resulted in it becoming as vulnerable as it is in England; its precarious state is simply the result of mixed fortunes. It is the plant's bad luck that it is tasty and slow growing, and therefore unable to survive in a landscape that is almost universally grazed. Centuries of sheep grazing in the Lake District have pushed many palatable species out of accessible areas into precarious and lonely refuges on cliffs and crags or, where these aren't available, into oblivion. It's a sliver of good luck that *Ajuga* ended up in one of these craggy refuges, and that botanists discovered and described it. Without this tiny helping of fortitude, it would be gone completely, and we'd never have thought of it as an English species. It's a safe bet that many other plants were pushed into extinction before we had the chance to record them.

I expect that few walkers in the Lakes notice the floral paucity under their feet; most will be unaware that things could be so much richer.

&

According to the *Oxford English Dictionary* a mountain is 'a natural elevation of the earth's surface rising more or less abruptly from the surrounding level and attaining an altitude which, relative to the adjacent elevation, is impressive or notable'. In the UK, an altitude of 2,000ft (610m) is used to distinguish between a hill and a mountain. I never in my wildest dreams thought that I'd end up with a job that involved the care of mountains, but using those definitions, several fall within our land at Haweswater. From the rounded grassy tops of Selside (655m) and Racecourse Hill (828m), to the craggy gullies of Harter Fell (778m), our mountains are as varied as they are harsh.

Exposed to the extremes of weather, mountains often have thin, poor soil, thanks to frequent rainfall washing away their nutrients. Snow cover means that plants growing in the mountains need to have a short growing season and to be able to flower and set seed much faster than lowland species. *Alpine* is the loose term given to this highly specialized group, regardless of whether the plants grow in the Alps.

Because soils are so thin, few alpine plants get the opportunity to put down deep roots to tap into subterranean stores of water, so despite the frequent rain, they must be drought tolerant. Species like roseroot and yellow mountain saxifrage, which grow along many of our higher-altitude stream sides, have solved this problem by having thick, fleshy leaves that make them look better suited to deserts than rain-soaked hills.

Water stored in their succulent tissues sees them through periods when the streams run dry.

Alpine plants often find themselves thinly spread across a wide area, reducing the likelihood of insects transferring pollen from one plant to the next. Viviparous fescue, a mountain grass, has got around this problem by ditching the need for pollination altogether. Where a typical grass would have a flower head, it instead produces a cluster of tiny new grass plants, genetic clones of their parents. By late summer, the viviparous fescue plants that grow profusely on and around Harter Fell look spectacularly punk, nodding under the weight of their babies.

In order to thrive in the cold, species such as pyramidal bugle and alpine saw-wort, among many others, are hairy. These hairs work in the same way as do those on our bodies, trapping what little heat there is near their leaves and stems, heat which helps them to grow faster. Other species, such as moss campion and purple saxifrage, have developed a squat, cushion-like form that retains both heat and moisture. By growing low, and with strong, lateral roots, they are able to eke out a living on virtually bare rock.

Purple saxifrage is my botanical nemesis. When I started at Haweswater, I was presented with a sheet of paper containing a list of twenty-six plant species, which had been written by Geoffrey Halliday, another member of Cumbria's botanical intelligentsia. Geoffrey is the lead author of *A Flora of Cumbria*, an incredible compendium that profiles the county's 1,373 plant species and presents maps of their locations, as surveyed by an army of skilled volunteers. The species on my list were ones that Geoffrey felt were particularly important for Haweswater, due to their rarity or notable habitat requirements. Virtually all are alpine plants and I set myself the challenge of finding them.

Some, like juniper, are widespread enough at Haweswater that no effort was required in ticking them off. For the others, most of which are hidden away in remote mountainous corners, I needed to put in more legwork. I had decent directions for a few and visited some with botanists who'd seen them previously. I've found bird's-eye primrose, alpine enchanter's nightshade, mountain everlasting and holly fern in several new localities, as well as in the places where they'd originally been discovered. Alpine saw-wort, which hadn't been seen since 1975, was refound by Caz, one of our volunteers with a keen eye for finding rare plants and lichens. I was only slightly sore that she beat me to it.

Purple saxifrage, however, remains stubbornly elusive. It is a plant with tiny thyme-like leaves, which can be completely obscured by its disproportionately large purple flowers. These blooms appear in early spring, before almost every other mountain flower. In the UK, purple saxifrage generally behaves as a chasmophyte, meaning that it grows in rock cracks. In landscapes with less grazing pressure, it's perfectly happy growing on more horizontal terrain, although being so small it is easily shaded and outcompeted by bigger species. The plant's stature also means it's virtually impossible to find when not in flower, so for the past few years I've made special trips in April in the hope of finding it. Derek Ratcliffe was the last person to see it at Haweswater, and it's his slightly hazy record from the 1970s that features on my list, but nobody has seen purple saxifrage here since.

I never met Derek, who passed away in 2005, though I'm familiar with lots of people who knew him well. He rates as one of the most gifted and influential naturalists that the UK has ever known. His ability to find and identify plant species verged on the superhuman. He wrote countless books on natural history and played a key role in setting up the govern-ment agencies responsible for protecting wildlife. If I could rediscover the purple saxifrage, it would feel like honouring his memory, a tiny testament to the resilience of nature, which he committed his life to protecting. There's no doubt that his purple saxifrage record is genuine, but rather frustratingly, Derek only recorded the general location for it: somewhere in the vicinity of Blea Water.

Blea Water is the deepest tarn in the Lake District, its dark waters plunging to an icy 63 metres. Shaped like a teardrop, it lies in a dramatic glacial corrie on the rugged western side of Mardale Common, just south of Riggindale and Sale Pot. Like some of the rocks on Harter Fell, the steep corrie walls that surround Blea Water are rich in calcium, making this another one of our hotspots for alpine plants. I've combed the corrie's nooks and crannies for purple saxifrage, to no avail, but I refuse to believe it's gone. The knowledge that this tiny precious plant might still be lurking somewhere is like an itch that I can't scratch.

Alpine plants in the Lake District exist on the very edge of a climatic envelope in which they can survive. Mountain loca-tions such as the eastern coves of Helvellyn, Honister Crag and Wasdale Screes, with their peculiar mixture of altitude, geol-ogy and aspect, support the majority of the Lake District's rarest alpine flowers, but other areas, including parts of the ranges near Haweswater, are also important. These high-altitude refuges can be thought of as islands, high enough to ensure that the plants growing here are free from competition with lowland plants, allowing them to cling on to life.

Alpine plants had their brief moment in the sun following the end of the last ice age, about 10,000 years ago. Bare land left in the wake of the glaciers would have favoured alpine plants over the more demanding, taller species, which rely on richer, deeper soils. As nutrients and soils accumulated, so did the vegetation, culminating in peak forest cover around 7,000 years ago. This would have resulted in alpines being pushed up into the hills and mountains, or in some cases to coastal cliffs where a replica of the post-glacial conditions remained.

Broadly speaking, that's where we still find alpine plants today, though humans have made their life an awful lot harder. The Neolithic period was when we first started having a major impact on habitats in the UK. Starting around 6,000 years ago, armed only with stone tools and our ingenuity, we cleared vast swathes of woodland and began the serious business of settled agriculture. We've more or less continued in that vein, exploiting and modifying the environment to suit our needs. Already restricted, palatable alpine plants would have been squeezed further by livestock grazing, pushing many into inaccessible refuges, like the crags around Blea Water.

Their adaptations to an extreme environment mean that alpines are a hardy group, which, combined with the fact that they are frequently beautiful, has made them popular with gardeners for centuries. Dorothy Wordsworth's *Grasmere Journal* recounts her plant-collecting forays into the Lakeland fells in 1800, bringing home 'her load' to transplant into the garden of Dove Cottage, where she lived with brother William. Casual collecting like Dorothy's probably only did a little harm, but things took a turn for the worse in the Victorian period, when alpine gardening became the height of fashion. The growing demand was met by intrepid plant collectors, who would scale even the remotest crags in search of their botanical quarry.

While this period of history gave a huge boost to the concept of decorative gardening, it also had a massive impact on the populations of plants growing in the wild. The insatiable appetite for orchids and ferns in particular – two groups that are slow growing and have complex ecological requirements – resulted in local extinctions up and down the country.

It's hard to measure the impact that the Victorian period had on wild plants, as there is scant information about their abundance in any historical record. For alpine plants, never common and restricted to the small patches of our only-just-alpine conditions, collecting almost certainly took a significant toll. While the theft of plants from the wild is now largely a thing of the past, as my pyramidal bugle visit shows, it hasn't died out altogether. Lately, our impacts have become even more pervasive, with both climate change and widespread nitrogen deposition resulting from industrial and agricultural activities shifting the delicate balance that the alpines rely on for survival.

It's tough being an alpine plant, especially in the UK. Perhaps because our native alpines are so restricted, they have a particular allure among botanists, myself included. You have to work hard to find them – I still haven't found the purple saxifrage – and they are often the focus for specific protection and conservation effort. This has warped our view of them, and it's easy to fall into the trap of thinking that the places we find our alpine plants are where they want to be. It was only by visiting upland landscapes elsewhere in the world that I realized how far from the truth this was.

❧

At a time when the natural world is collapsing around us, perhaps you think that worrying about small, fussy alpine plants is a distraction. But, as species on the front line of the climate catastrophe and being so sensitive to many other widespread environmental problems, they carry a significance that far exceeds their stature. And sadly, it isn't only plants in the mountains that are struggling.

In 2012, the charity Plantlife produced a thoroughly disheartening league table of plant extinctions. It showed how many species had been lost per county over the course of the twentieth century. At the top of the table was Banffshire in Scotland, with a score of 0.9 species that had been lost on average from the county per year. In coveted last place was Wiltshire, with a much less frightening figure of 0.08 – but that still equates to an extinction every twelve years. Cumbria hovers somewhere mid-table at 0.37, meaning that we have lost one species roughly every third year. That's fifteen plant species that have vanished from the county in my lifetime. The Cumbrian losses include burnt orchid and lady's slipper orchid, conspicuous and coveted flowers that would not have been overlooked if they were still here.

This litany of loss tells of dwindling diversity, but wildflower abundance is equally important, if not more so. Many species of wildflower can persist in tiny patches for years, but this doesn't mean that they wouldn't rather be all over the place. But, for a variety of reasons, huge swathes of the countryside have become flower-free zones. Not only is this a huge loss for the many bees, butterflies and other insect lives that depend on them, but it also means that the chances of having flower-rich experiences have been taken from us.

Because many of my RSPB colleagues are such ornithological experts, since starting at Haweswater I have focused my energies on wildlife of the unfeathered kind and have been growing ever more botanically minded. Flowers are every bit as engaging as birds. As well as having fascinating stories and cultural associations, flowers have the added benefit of not flying away when you want to study them more closely.

As someone whose wildlife radar is always on, it often surprises me how many of the details of nature, and flowers in particular, that seem so conspicuous to me are overlooked by others. Perhaps I shouldn't be so surprised. As we become an increasingly urban species, our exposure to nature is diminishing. The wonderful wildlife documentaries that provide valuable proxy connections to the natural world focus almost exclusively on animal life; the plants are there, but they are rarely named or given much attention. Yet without these plants to eat, hide behind and nest in, the elephants, tigers and birds-of-paradise simply wouldn't exist. We've not been taking good care of these life-giving foundations.

❧

I live at the top of a windy hill on the border of the Lake District National Park with views east to the Pennines, west over Blencathra, and south towards Great Mell Fell and the central Lake District beyond. On a clear day I can see the long ridge of High Street from my door, marking the edge of our land at Haweswater, and I can see back to the house on the days when I'm working up there. It takes me a little over half an hour to drive to work. It's only eleven miles as the crow flies, but straight lines don't get you far with lakes and mountains in the way.

High Street divides the two poles of my life. At Haweswater I'm bending the land to new purposes, helping it to become richer and wilder. At home, things are different. Our garden backs onto the highest dairy farm in the country, land over which I have no say. It was as much the observations of nature on my doorstep as those at Haweswater that opened my eyes to the enormous scale of the challenges facing our wildlife.

Over the years, I've come to know the scruffy verges that flank our quiet road intimately, having driven and walked alongside them thousands of times. To some, they probably don't warrant a second look, but in spring and summer they become narrow strips of wonder, teeming with wildflowers. Water avens is the most conspicuous member of my local verge community. A member of the rose family, its scruffy pink and yellow blooms are a magnet for bumblebees. In late summer, the drooping flowers develop into spiked seed heads, looking like medieval maces, protecting the verge against intruders. Here also is knapweed, meadowsweet, devil's-bit scabious, wood and meadow crane's-bill, bush and tufted vetch, bistort, valerian, common twayblade, heath spotted and early purple orchids; a wonderful diversity of bold, attractive flowers providing nectar to insects and colour and beauty to us.

On the other side of the hedge, things are not so cheery. These are conventional dairy fields, where it's rare to find a flower of any kind amid the bright green grass. Through a cycle of spraying, ploughing, reseeding and fertilizing, the fields are maintained as high-energy rye-grass monocultures, cropped several times a year to make silage. In the current market-driven food and farming industry, this intensity is a necessity for survival. The biological past of these fields has been comprehensively erased, with only the verges now giving a clue to how they once were.

One summer evening, while walking along our road, I watched a barn owl ghosting over one of the verges, glowing in the failing light. It seemed uncomfortable, its flight peculiar, wobbling like a tightrope walker as it hunted for small mammals or frogs. The tangle of plant life in the verge gave both food and cover to sustain the owl's quarry, as well as sustaining the insects, molluscs and worms that some of this quarry would be hunting for in turn. The verge was a tiny but functioning ecosystem, with an intact food chain that relied on healthy and diverse plant life.

Forced into the indignity of hunting like this, pushed into the only sliver of land that had a hope of harbouring its prey, the owl is a striking visual symptom of a wider malaise. Roadside verges create a blur of colour through our car windows as we speed past. They are the most frequently seen bits of the countryside, but they are a thin veneer that all too often screens a far bleaker scene beyond. A deflowering has happened across vast tracts of our countryside, both in the intensive farmland and the seemingly wilder hills above. Over time, we've come to accept a landscape in which wildflowers survive only in the liminal spaces as normal. Yet without them, the rest of life has little chance.

Because this has been playing out over many decades, few people can recall how things used to be – our baselines have well and truly shifted. Things have got so bad that many people aren't even able to appreciate flowers when they are in full view. We are becoming plant blind. Two American botanists, James Wandersee and Elisabeth Schussler, came up with the term plant blindness in 1998, which they defined as:

a) the inability to recognize the importance of plants in the biosphere, and in human affairs;
b) the inability to appreciate the aesthetic and unique biological features of the life forms belonging to the Plant Kingdom; and

c) the misguided, anthropocentric ranking of plants as inferior to animals, leading to the erroneous conclusion that they are unworthy of human consideration.

Wandersee and Schussler theorized that the basic biology of our senses is partly responsible for plant blindness. Our eyes capture far more information than our brains are able to process, so everything that we see is selective. Unless objects that our eyes see already have a particular resonance, they simply blend into the background. As flowers become less of a feature in our everyday lives, our ability to ascribe meaning or memories to them diminishes, and so we literally notice them less. The less we see them, the less likely we are to notice them vanishing, or to do anything to stem the losses.

Happily, we can train our brains to claw their way out of this downward spiral. Buying a wildflower guide, taking a walk with a friendly naturalist or going on a guided walk is a sure-fire way to train your botanical perceptions. Learn how to tell a bluebell from a harebell and you'll start to notice all the things that aren't either. Even if you don't have the type of memory to recall the names of things, seconds spent appreciating the delicate perfection of an eyebright flower or the leopard-like leaves of a common spotted orchid will never be wasted. Invest some time in this and you'll soon find that you can't stop noticing the flowers – and you'll miss them when they're not there.

The stark contrast between verge and field reminded me of the scale of the challenge. In order to get our eagles back, we were going to have to start with the flowers. In order to look up, we must first look down. I didn't anticipate it, but a focus on flowers was to put me at odds with an influential slice of Cumbrian society.

CHAPTER 3

Locking Horns

HAWESWATER:
Hafr's Lake
(Old Norse, Hafr being a personal name)

Six months after my arrival at Haweswater, my colleague Bill and I were invited to Dalemain House, seat of the Hasell family since 1679, and the nerve centre of the Dalemain Estate. Although it was an important opportunity to communicate with our neighbours, due to my inexperience and uncertainty in my new role it felt at the time like a summons.

A good chunk of the Ullswater Valley falls within the Dalemain Estate, including the western side of the High Street ridge. This makes Dalemain our neighbours along the watershed boundary of Bampton Common, the largest of the three commons that our Haweswater tenancies give us grazing rights over.

Famed for its annual marmalade festival, quirky medieval tea room and gentle parkland scenery, Dalemain House and its glorious gardens are all grandeur and pleasantry. It's the most conspicuous part of the Estate, a pale pink mansion set back from the road between our local town of Penrith and Ullswater.

Ullswater is Haweswater's more famous twin, lying in its valley five miles to the north-west as the crow flies, or fifteen miles by road. It is one of the Lake District's largest and most popular lakes, and justifiably so. The campsites and cafés, the steamer that runs up and down it and the many people who come to walk its miles of footpaths might take a toll on its tranquillity, but the Ullswater Valley remains a spectacular place, a perfect postcard of soaring peaks and wooded crags, reflected in the deep water.

Ullswater is important for me personally, too. I married my wife Becki in a stone barn a mile and a half from its lapping shore on a still August morning. As our guests arrived – the rural crowd in wellies, our urban friends wearing carrier bags over their smart shoes – a fine summer rain fell. Being 370 metres above sea level, we prefer to think that we were married in a cloud. Our children, Elliot and Aphra, go to the village school at Ullswater's southern tip. Watching the valley change with the seasons makes the school run a daily joy.

The Dalemain meeting to which Bill and I had been called was to discuss the impact that our land management might have on the estate's tenant farmers, all of whom were also invited. As meetings go, it wasn't one I was looking forward to; the two of us and a room full of people we knew had deep reservations about us entering the frame. Because there is only a partial boundary between Dalemain-owned land and the land that our sheep graze around Haweswater, a shift in the way that we manage our livestock could have a knock-on effect on their farming. There were also ideological concerns. How might a large NGO like the RSPB fit into a scene which for generations had been the sole preserve of farmers?

Bill has been an RSPB warden at Haweswater for most of his working life and was instrumental in securing our farm tenancies. Tall and lean, he's hard to keep up with out on the fells. It was the eagles that first lured him to Haweswater and his piercing gaze and unwavering focus give him a resemblance to the birds he's spent so many years serving. He's unflappable under pressure and so, while I panicked about the meeting, he provided calm assurance that all would be well.

Having an idea of what my audience's concerns might be, I tried to address as many as I could in the presentation I delivered in Dalemain's oak-panelled meeting room, before they were raised. We would be increasing employment, I emphasized. We would still have plenty of sheep, would respect our neighbours and put measures in place to limit impacts on them. Being so new in the job, my presentation was nervously delivered and probably too full of conservation jargon. Nonetheless, I hoped I'd done a decent job. The reaction of my audience told me otherwise.

As the farmers began putting up their hands, we were accused of dismantling thousands of years of custom and practice, of land abandonment, of rural depopulation, of not being neighbourly, of removing a stock of sheep that could trace their ancestors to the time of the Vikings, of reducing the nation's ability to feed itself, and of wasting money. Evidently my presentation hadn't communicated that we weren't guilty of any of this as clearly as I'd hoped.

One of the most vociferous contributors was an aged farmer who had spent my entire presentation fast asleep, apparently waking up with the express intent of venting as much bile in my direction as possible. It felt that everyone in the room had come believing that we were about to destroy everything they held dear, and went home thinking in exactly the same way.

I left feeling so bruised and shaken that, as I sat in the car outside, Becki had to talk me down over the phone before I was in a fit state to drive home. The strength of feeling that had been expressed made me wonder: were we doing the right things at Haweswater? Did we really have the right to tinker with a farming system that was considered so sacrosanct? I was starting to realize that I had a lot more thinking and learning to do, and that in order to survive in this new job I would need a thicker skin.

It might seem strange that the Royal Society for the Protection of Birds would end up running a pair of hill farms complete with flocks of sheep. In fact, we were already running many different farming operations across the UK, from an arable farm in Cambridgeshire where techniques have been developed to help wildlife survive in intensively farmed landscapes, to nature reserves grazed by sheep, cattle, goats, pigs and ponies. What we didn't have, until Haweswater, was a detailed understanding of hill farming. Haweswater is giving us the chance to learn if and how hill farming can help to boost populations of alpine plants, insects, birds and other wild creatures, while also improving water quality, carbon storage, flood-risk reduction and other so-called public goods.

The bulk of the land in our care at Haweswater is common land. A common is a piece of land owned by one person or organization, but over which others have certain rights. Historically, the commons were the poorest bits of large estates, sometimes referred to as manorial wastes, which peasants were given permission to use. Over time these customary uses became enshrined in law, with rights of common becoming attached to specific property, and only the owner or tenant of those properties being able to exercise the rights. Common land isn't therefore land that just anyone has a right over, as is often thought.

Historically, there were many widely practised rights of common. The right of turbary allowed the cutting of peat or turf. The right of pannage, to graze pigs on acorns. The right of estovers, to take firewood. Many of these rights still exist in the deeds of farms and in commons registers, but for the majority of upland commons in the Lake District, where trees are few and far between, the only right worth exercising today is the right to graze livestock.

Managing livestock on common land, or commoning, is a skilled and challenging business, relying on cooperation and collaboration between commoners. Although ponies and cattle are grazed here and there, it is sheep that dominate the commons of the Lake District. Sheep are partially territorial, and if kept to a particular piece of ground by shepherding, they become habituated to it. Each year they lead their offspring back to the same area, passing their geographical bond down the generations. Farming sheep in this way is known as 'hefting', with each flock's traditional area known as a heft.

Because there are no physical boundaries between hefts, the system operates in a loose equilibrium, relying on sheep being at a more or less equal density across the hill to keep them where they are supposed to be. If the number of sheep in one heft reduces, then more grass will be able to grow, tempting sheep from the neighbouring heft to move in and take advantage of the extra bite. The gathering of livestock from commons is carried out communally by several commoners and their dogs at various points throughout the year, and any stray sheep, once identified by their paint marks, horn brands or ear notches, are swapped back to the right flock.

We're involved with three of Haweswater's commons –
Mardale, Bampton, and Rosgill and Ralfland. Our co-commoners
have every right to know about our plans, just as we need to
know about theirs. Nobody has a free hand on common land,
and livelihoods depend on getting on with the neighbours.

&

I regularly catch myself referring to Naddle and Swindale as
'my' farms. People often speak of places as being theirs, but I
think perhaps the opposite is true. Whichever way it runs,
we've become entwined. As I've poured myself into it, some
essence of the land I look after has seeped back into me.

Land ownership is at the root of many of the challenges
facing the countryside. Although land is generally privately
owned, the services that flow from it – the purification of air
and water, the regulation of the climate, the wildlife – are of
public benefit. Ownership becomes an even thornier issue in a
place like the Lake District which, 140 years before it became
a national park, William Wordsworth – the Lake District's
most famous son – described as 'a sort of national property in
which every man has a right and an interest who has an eye to
perceive and a heart to enjoy'.

Unlike national parks in many other parts of the world,
those in the UK are not owned by the state. They are held
instead by individuals, public and private organizations and
charities, no different to any other part of the countryside.
That doesn't stop many people feeling like they have a greater
stake in the place. Public interest adds a third dimension to the
already complicated, age-old tussle between landlords and
farming tenants.

Like many farms, Naddle and Swindale are rented; United

Utilities are our landlord. The Haweswater Estate that they own incorporates eight farms and several large blocks of open fell land, and all the woodland, bog, grassland, rivers, tarns, cliffs and crags therein. Before the RSPB got involved, Naddle and Swindale had a series of farming tenants stretching back into history, and each would have felt their own strong sense of ownership.

We are lucky at Haweswater; our aims are closely aligned with those of our landlord. Although we pay them a rent, we are also partners, working towards the restoration of the ecosystem. Doing this will make the land act as a better filter, helping purify United Utilities' water between rainfall and reservoir. The RSPB wants the land to act as a better habitat for wildlife. Fortunately, the same interventions deliver for both ambitions.

However, not all tenants see eye to eye with their landlords. The National Trust owns more than 20 per cent of the land in the Lake District, which includes more than ninety tenant farms. The arrangements they have with their tenants vary in duration and detail, depending on when they were signed up and what the Trust's interests were at the time. The National Trust have recently increased their ambitions for the natural environment. They want many of the same things for their land that we want for ours at Haweswater. This doesn't sit easily with all their farming tenants, many of whom see livestock production as their primary purpose. This isn't to say that they don't want to help nature too, but for most, it isn't the main reason they started out in farming.

A friend who used to work as a land agent for the National Trust had the unenviable job of trying to square this circle. On a visit to one of her farming tenants, one with a long-term, three-generation tenancy, things got ugly. The farm was the only home this farmer had ever known, and he probably felt like it was at least as much his as it was the Trust's. My friend's suggestions of new ideas for nature-friendly farming were taken as a criticism of what had come before and tensions flared. Verbal abuse escalated to threats of violence, and eventually she had to push him away before she could get into her car and escape. She stopped visiting the farm by herself. It's an extreme example, but it demonstrates the strength of feeling that exists in hill farming, as well as the resistance to change in some quarters. In this emotionally charged environment, farming for nature in the Lake District means treading with care.

Even farmers who own their land are still not masters of their destiny. In the uplands, where the climate and conditions make farming an economically marginal activity, public funding is the only thing that really keeps farm businesses going. A typical Lake District farm will derive a significant proportion, if not the majority, of its income from government support payments so that to a degree, farmers are beholden to the state and the taxpayer. I'd be surprised if there were any farms in the Lake District that weren't in receipt of government support, making Wordsworth's definition of the Lakes as 'a sort of national property' truer now than ever.

The bulk of the day-to-day care of our livestock is carried out by Matthew, our farm contractor. Broad-shouldered and built for an outdoor life of hard physical graft, he is both knowledgeable and steadfast. He studied farming at Newton Rigg, the local agricultural college, but I suspect he learned more than the college ever taught him by helping run his grand-father's farm in the shadow of Blencathra, the saddle-backed mountain that dominates the skyline twelve miles north of Haweswater.

Matthew started working for us in 2015, when he was twenty-eight. Coming from a conventional farming background, he's always retained a degree of scepticism about the approach that we're taking. He thinks we need more sheep, I think we need less, but we're always able to talk things through and come up with workable compromises.

Working for us isn't easy, but Matthew has learned to ignore the gossip that bubbles to the surface from time to time. We get along well, and I've learned a huge amount from him over the years. His skill, and the innumerable decisions he has taken about breeding and feeding, buying and selling, have slowly and surely improved the quality of our flock.

We keep Swaledale and North Country Cheviot sheep, and regularly achieve top prices. Swaledales are the taller, leggier breed of the two. A Swaledale ram is an impressive curly-horned beast, stylized in the Yorkshire Dales National Park logo. Cheviots originate in the hill range after which they are named, spanning northern Northumberland and the Scottish Borders. For us they are a more reliable breed than the Swaledales, less prone to jumping fences and walls, which makes them consider-ably less of a headache.

Our sheep are sold at one of the local auction marts – busy, noisy places that are as much about social exchanges as about money. The selling is one of Matthew's jobs and should be the highlight of the farming calendar when the year's efforts are judged and valued. At the mart, Matthew often finds himself chatting with potential buyers who are full of compliments for our stock – right up until the moment when they learn that these fit, healthy sheep have come from a farm run by the RSPB. At that point, they often walk away muttering about not wanting to buy from 'those arseholes'. I don't get involved. I suspect that my presence at the mart would drive the price of our sheep down even further.

꧁

Although there are many native *breeds* in the UK, such as the Swaledales and Cheviots that we farm, sheep are not a native *species*, and this is an important distinction. The wild ancestors of sheep came from Mesopotamia and have been domesticated for at least 6,000 years; there were no sheep in the pre-human British countryside. A native breed is simply one that has been bred in a particular area for traits that help it to thrive in the local conditions. Native species, on the other hand, have a more ancient connection to a place and the other native species they share it with. Assemblages of native species are associated in a myriad complex and subtle ways, natural selection having weeded out those that weren't able to get along.

Evolving alongside the ancestors of pigs, cattle and ponies, our native plant life has adapted to cope with the particular way in which these species graze and browse. Sheep, on the other hand, graze very differently. Their small mouths allow them to pick out sweeter, more delicate species, ignoring coarser, less palatable ones, which end up dominating.

The few thousand years that sheep have been in the UK is nothing in evolutionary time, not nearly long enough to have allowed our native flora to adapt to their presence. This is why, in addition to their often excessive numbers, sheep have had such a serious impact on our ecology compared with other grazers, and why our aim of improving the state of Hawes-water's ecology while keeping sheep is such a challenge.

❧

On a traditional hill farm, lambs are born in spring, five months after the ewes have been put to the ram, referred to locally as the tup. Lambing usually takes place as close to the farmyard as possible so that the farmer can keep a close eye on things and try to reduce mortality. After a few weeks, the ewes with single lambs will be turned out to the fell, while those that have had twins will be kept near the farm for longer to build up their strength. By mid-summer, the whole flock will be up on the fell apart from the tups, left in their bachelor group, longing for the return of the amorous autumn. Putting the sheep up onto the higher ground gives the more productive, enclosed land on the valley bottom a break, some of which will have livestock completely shut out for several months to allow a crop of flower-rich hay to be grown. This hay will provide the fodder to see the sheep through the winter months, after which the whole cycle begins again. This is broadly how we've done things at Haweswater since taking over the tenancies, though we run considerably fewer sheep than comparable farms.

You might wonder why we keep sheep at all. From an eco-logical point of view, they contribute very little, and if we were running Haweswater as a straightforward nature reserve, we probably wouldn't have any, replacing them with a scattering of hardy cattle and ponies, and increasing the areas with no grazing at all. While this approach would undoubtedly be bet-ter for wildlife, we would lose the opportunity to influence others. The point is to try to show that a thriving natural envi-ronment can not only live alongside a farming operation, but perhaps even help it to thrive economically. We might then inspire the widespread restoration of a harmony between nature and farming that has become discordant.

Whether we like it or not, new models of farming are going to be needed to weather the many storms that are approaching, be they climatic, ecological, political or economic. The greatest is undoubtedly climate breakdown, which is no longer some remote prospect, but a real and present danger. There have been some momentous climatic events since I've been at Haweswater, such as record-breaking storms and floods, summer droughts and winter snows. The setting of new extreme temperature and rainfall records has become so commonplace as to now seem unremarkable.

At the same time, global biodiversity loss has continued to accelerate. When I was growing up in Devon, we had a buddleia bush in our garden. I have vivid memories of it being so smothered with butterflies that it seemed about ready to take flight. Buddleia isn't a native species to the UK, having been brought over from China in the 1890s to brighten up Victorian gardens. Its light, winged seeds soon made a bid for freedom, and the species is now widely established in waste ground and often forms conspicuous stands along railway lines. Sometimes referred to as 'butterfly bush', its long cones of fragrant purple flowers are irresistible to both butterflies and moths in the summer months.

The buddleia in my parents' garden persisted long after I flew the nest. I'd see it most years on visits. Despite growing older and leggier, it always managed to put on a good show of flowers, but with each passing year there seemed to be fewer and fewer butterflies feasting on them. It's easy to pass off these sorts of recollections as rose tinting, as a trick our memories are playing on us, heightening past experiences, brightening the colours. The sad truth is that my memories of that buddleia and its dwindling insect life are probably completely accurate. All the evidence suggests that there actually were more butterflies, birds and bees in the garden of my childhood.

Quantifying the decline of nature in our countryside isn't an easy task, but the State of Nature report, first published in 2013, did just that. It was the product of the efforts of over fifty nature conservation organizations and provides the most comprehensive account of UK wildlife ever produced. It has been updated twice since its original publication. None of its versions make for happy reading.

The State of Nature report provided data for a huge range of species and habitats, and although some are coping better than others, the headline is simple: nature in the UK is in free fall and we are rapidly losing both the diversity and abundance of our wildlife. There are now 44 million fewer birds in the UK than fifty years ago. Some 15 per cent of our species are at risk of extinction, and 2 per cent have already gone the way of the dodo since 1970. We've lost 97 per cent of wildflower meadows since the Second World War, 94 per cent of our turtle doves since 1995, half of our hedgehogs since 2000 . . . and the list just goes on and on.

Perhaps the most depressing thing about the State of Nature report is that its baseline, against which all the declines are measured, is 1970. By this point in history, wildlife had already suffered twenty years of intensive post-war agriculture, two centuries of pollution resulting from the Industrial Revolution and the widespread persecution of wildlife that was a prominent feature of the Victorian period. Things would have already been pretty bad by 1970. All the State of Nature reports show is how much further things have slid since. I know this is a gloomy story that you've heard before, but thankfully the level of concern now seems to be commensurate with the enormous scale of the crisis.

The reasons for these devastating changes to nature are manifold. Climate breakdown, urbanization, pollution, land drainage, invasive non-native species, habitat fragmentation, disturbance, disease, over-exploitation and hunting all play a part, and in many cases work together to make life unbearable for wildlife. The single most important factor, however, is intensive agriculture. This is an uncomfortable truth – particularly if you're a farmer.

Farming is a broad church. The intensive arable farms of the lowlands, over which gigantic, computer-operated harvesting machines lumber, are a far cry from the pastoral farms of the Lake District. What they have in common is that for more than half a century all farms have been financially incentivized to increase their production.

Farmers have always done what society has asked of them. For the past seventy years, the overriding demand has been to produce as much food as possible. Farmers responded and their success has been extraordinary. Since the 1970s, using the full range of technologies available, agricultural productivity has risen by 150 per cent. That's pretty remarkable when you consider that it has been achieved using roughly the same area of land. But it came at a cost. During that same period the farmland bird index fell by 54 per cent, tens of thousands of miles of hedges were removed, and pesticides, fertilizers, antibiotics and slurry ended up in our watercourses, periodically killing off aquatic life. We all need to eat, and we all like our food to be cheap, so this is a problem for all of us.

Farming has been a big part of the problem, but unlike all the other factors driving wildlife decline, farming is unique in that it can also be a big part of the solution. Sensitively farmed landscapes can be some of the most nature-rich places in the temperate world, but finding the right balance between feeding ourselves and helping wildlife to thrive is no small challenge.

Over the last few years, there has been a long-overdue national awakening. The entangled nature and climate crises present genuine threats to the way that we live our lives. It is this gloomy context, coupled with our departure from the EU and the development of new government policies, that has resulted in the public discourse around farming and land use reaching fever pitch. As the UK's largest nature conservation charity, it's hardly surprising that the RSPB is responding – it is our duty to do so – and we've found ourselves increasingly in the spotlight. At Haweswater we have been trying out new ideas in the hope that they might be replicable elsewhere. We have explored ways to reduce flooding, reduce plastic and fossil fuel use, increase how much carbon the landscape could lock up and understand what role livestock farming could play in all of this. It feels like we're doing the right things, but not everyone approves.

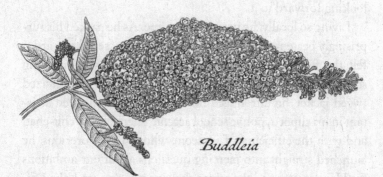

Buddleia

A month after the Dalemain meeting, John, my closest friend at United Utilities, and I were visited by our local MP, Rory Stewart. John's good fun, irreverent and witty. He did a stint with Cumbria Wildlife Trust, where I'd also worked before starting at Haweswater, and so we know a lot of the same people and places. John has a longer connection to Cumbria than I do, having moved to the county when he was three. Standing in the weak October sunshine next to our recently refurbished sheep pens at Naddle Farm, John in his green United Utilities uniform, and me in RSPB blue, we discussed what we should show Rory while we waited for him to arrive. He was only able to spare a couple of hours, and we were keen to make a good impression.

Rory had been my constituency MP for about three years, and I'd been impressed with how approachable he seemed to be, and how well he'd stood up for important local issues. As he lived only a few miles north of Naddle Farm, I drove past him while he was out running from time to time, but this would be the first occasion on which I'd spoken to him. I was looking forward to it.

Living so locally, he was right on time. As he parked his surprisingly beaten-up old car in our cracked concrete farmyard, I felt the familiar twinge of nerves that comes with meeting someone of such high social standing. Pulling on an oversized tweed jacket, he introduced himself and his fresh-faced assistant in his clipped, public school accent. There was no chit-chat, and with the efficiency of someone whose time is precious, he launched straight into piercing questions about our ambitions for Haweswater and the wider challenges facing the Lake District as we set off up the stony track into Naddle Valley.

Rory is a fiercely intelligent man with a deep connection to the land. While we were with him, he told us that if the Lake District were to go the way of the Scottish Highlands – by which I assume he was referring to the Highland Clearances of the eighteenth and nineteenth centuries, when tens of thousands of rural people were thrown off their land by profiteering landowners – that he would kill himself. He floated interesting ideas about how the national park could be divided into zones with different emphases – one for traditional pastoral farming, another for nature and wildness, another for recreation and so on. We talked him through our nature-friendly farming system, our plans for scattered tree planting, hay-meadow and bog restoration, and how we believed it could all be done in sympathy with the Lake District's farming traditions.

Waving him off after posing for an awkward photo back in the yard, John and I felt like we'd done a decent job in getting our points across. Rory's blog post and column in the local paper a week later seemed to back this up, describing the visit and the 'fantastic projects' he'd learned about.

Three months later, Rory apparently had a change of heart. A new article appeared on his website describing the same visit that he'd written about so positively, but this time in completely different terms. For some reason he ignored both the verbal and written information that John and I had provided. His article claimed that we showed him an area 'of eight hundred hectares, which was being planted as forestry'. That's more than the entire area of enclosed land within our tenancy, and we've never had any plans for forestry. He suggested that we were trying to create a 'landscape which had not existed almost since the first human settlement', when we had a flock of around 1,500 sheep, were constantly maintaining our drystone walls and hedges and restoring one of the largest suites of hay meadows in the national park. He told his readers that he didn't see any sheep, but that was because we were in the valley bottom and, being autumn, they were all still up on the fell. He highlighted the difference between our farm contractor and someone from the 'indigenous population', saying that we shouldn't be 'bringing professionals in from other parts of Britain'.

Not only had Rory reinvented our conversation, but it also felt like he was poking fun at us. He referred to John and me as being dressed like 'canoe instructors', a reference that would probably have gone over the heads of most casual readers, but one that was pitched perfectly to accentuate a sense of rural tribalism, where conservationists are characterized as being distinct from farmers, and to make it clear on which side Rory's allegiance lay.

When John and I had recovered from our disappointment, we began to consider why Rory might have made such an about-turn. In a very rural constituency a bit of RSPB and United Utilities bashing might strengthen his support with what he clearly viewed as the most important part of his electorate: the farmers, and not people like me.

꩜

A year into the job, I felt isolated and embattled. What I saw as scars on the fells left by decades of intensive farming enraged me, and rather than blame the policies that had caused them, I quietly cursed those who had let them develop on their land. I didn't show it outwardly, but the more I learned about how diminished our wildlife was, the angrier I became.

The opposition felt formidable: the local MP, the neighbouring estate, the massed forces of Cumbria's farmers and their organizations. All seemed ranged against us, and by extension it felt like they were against me. There were a handful of people spreading deliberate untruths in order to discredit our work, or to get us to quit. One rumour that our livestock were all at death's door due to maltreatment was particularly hurtful, especially to our farming staff. I can see now that my inexperience had allowed these malicious rumours to proliferate. For someone who naturally has low self-esteem, it wasn't a happy place to be.

With a sense of powerlessness I tried to fight the rumours with facts, but the stress I felt during face-to-face meetings meant that I invariably shouted people down with rehearsed arguments, irrationally believing that would bring everyone around to our way of thinking. It didn't get me far. Feeling like I was failing to make any headway, I started to question whether it was worth me even taking part in the debate.

If I'd stopped to think, I could have reassured myself that there were plenty of people who shared our world view. The million-plus members of the RSPB were surely on our side, but it didn't feel like any of them were making their support known, and so I grew increasingly convinced that everyone was against us.

CHAPTER 4

Boom and Bust

HIGH STREET:
The high paved road
(anglicized from the original Latin name, Via Alta)

With opposition mounting, my resolve began to buckle. As much as I believed that what we were doing at Haweswater was right and important, I began to worry that I wasn't strong enough to take the constant criticism. I decided to seek professional help in the form of talking therapy to try to keep some dark thoughts at bay. I even started looking for other jobs and got as far as interviewing for a place on a teacher-training course. I'd had some experience of teaching while living overseas, and figured I'd probably manage.

The talking therapy slowly helped me to realize that part of why I was feeling so low was due to a lack of control. It felt like a lot of local society was against me, and that I didn't have any power to do anything about it. We had effectively been forced to backtrack on plans to remove sheep from a part of one of our commons, pressured into continuing to graze an area that we knew was being damaged. We'd received letters to senior managers, articles in the paper and frequent face-to-face criticism, and in the interests of avoiding a spat we had backed down.

So, I started running. That at least I could control. I'm not naturally athletic, so at the beginning it was hard, and I'd come home exhausted after a gruelling one or two miles. But slowly, two miles turned to three, then six, and after a few months I could be out for an hour and enjoy the feeling of adrenaline and of covering ground. Not only did the improved fitness help my mental health, but the sense that I had control over at least some aspects of my life helped my confidence to grow. I always listen to music while I run, and maybe 'Rage Against the Machine' helped too. The power in Zack de la Rocha's voice as he spits, 'Fuck you I won't do what you tell me!' strengthened my resolve, and certain faces floated into my mind's eye as I mouthed along, feet pounding the gravel. After six months of training, increasing my maximum distance by a couple of miles a month, I entered and completed the Haweswater half marathon – a fitting place to run my troubles away.

I can't recall whether it was the talking sessions or a thought that struck me while running, but it occurred to me that one of the things that I was lacking was a community. In the Lake District, the farming community is revered; many farmers can trace their origins back for centuries, and although frequently riddled by multi-generational family feuds, it really is a community united around a shared purpose – to produce food.

What, though, about the 'conservation community'? In Cumbria, the RSPB employs twenty people across five Nature Reserves supported by over 100 volunteers. Meanwhile, Cumbria Wildlife Trust have fifty staff, plus 500 active volunteers. The National Trust have over 300 people on the payroll in Cumbria, while Natural England and the Environment Agency employ a few hundred more. The Woodland Trust,

Rivers Trusts and a number of other smaller charities add in a further thirty or so. Then there are all those private individuals, naturalists, landowners and ordinary citizens who are passionate about the natural world. Like the farming community, this is another group with a shared passion and a shared purpose. They believe – we believe, I should say – that while food production is an obvious necessity, the land should be as much about providing for wildlife as it is about providing for us. There are of course also many people who have feet in both camps.

If I was having a hard time and feeling isolated, I figured that some of those hundreds of other people working in conservation in Cumbria might be feeling something similar. So, I started forging links with people doing similar jobs to me across the county. It felt uncomfortable at first, like asking someone out on a date, but the response was enthusiastic, and a few years later I'm part of a thriving network of conservation practitioners and nature-friendly farmers who meet regularly to discuss shared issues, view examples of good land management, and encourage and inspire each other. Knowing that I can tap into this wider constituency of people who think and feel as I do has definitely helped put things into perspective. I don't feel so isolated any more.

Although employed by other organizations, there are many people working in conservation in Cumbria who I think of as colleagues, and as friends. Without Simon and Jean at Natural England, John at United Utilities, Olly at the Environment Agency, Pete at the Woodland Trust and many others, working at Haweswater wouldn't be half as much fun, and we wouldn't have achieved anything like as much as we have.

Historically, sheep were not the be all and end all of farming in the Lake District. Getting a grasp on how their numbers have changed through time has been key to developing my understanding of the ecological history of the region.

For centuries, the number of animals that a farm could carry was determined by nature and common sense. There was only so much hay that could be grown on the low-lying land, and it was this that set the limit on how many mouths could be fed in the winter. Anyone seeking to push these limits took a massive risk – a hard winter with too many livestock could leave a farm with a whole load of deadstock.

Within these natural thresholds, livestock numbers rose and fell in relation to agricultural booms and depressions. From 1905, the story can be told with more clarity, thanks to government data from official farm surveys, which continue to this day.* For most of the first half of the twentieth century, up until the end of the Second World War, farming in the Lakes was similar in many ways to how it had been for centuries. Largely unmechanized, it relied on the sweat and muscle of a good chunk of the local community. The work was hard, and yields were low by modern standards, but most farms produced a broader range of livestock and arable crops than today. Alongside sheep, cows, pigs, goats, chickens and geese, most farms had orchards with apples and damsons, kitchen gardens and small arable plots, growing wheat and oats, turnips and

* All figures stated for Cumbria have been adjusted to account for the county boundary changes that occurred in 1974, when Cumberland, Westmorland and parts of Lancashire and the West Riding of Yorkshire were amalgamated to form the current county of Cumbria. The figures for the period before the boundary changes use the combined figures for Cumberland and Westmorland multiplied by a factor based on the difference between the size of these historic counties and the current size of Cumbria.

potatoes, and whatever else could reliably cope with the climate. In those halcyon days before the widespread use of fertilizers and pesticides, the produce was all organic and wholesome, and farming and wildlife were comfortable bedfellows.

Livestock numbers were fairly stable between 1905 and 1950, with the total number of sheep and lambs in Cumbria hovering around the 1.1 million mark, but big changes were on the way. With rationing still fresh in the national memory, the late 1940s and 1950s saw the beginning of a huge push to boost productivity, in part to insure against shortages should another war break out. By the 1960s, the government started to offer production-focused grants, paying for wetlands to be drained and for hedges to be removed to make farming more efficient, jobs that were made easier by the ever-growing power of farm machinery.

The focus initially was in the more productive lowlands. It was here that the most damage was inflicted, where the bulk of the hedgerows were removed and where resulting gains in yields were made in an astonishingly short time. The impact in the uplands, where land is harder to physically alter, were subtle at first, but as the financial incentives began to trickle in, drainage of peat bogs and other wetlands became commonplace. Linear scars cut by diggers into sodden hillsides sped the flow of water downhill, and ponds and pools vanished from the maps.

The number of sheep and lambs in Cumbria had climbed steadily from 1.1 million in the 1950s, to 1.5 million by 1975, a modest intensification of farming compared to more productive regions. Recognizing that the uplands were at risk of being left behind, the government introduced the Hill Livestock Compensatory Allowance in 1976, under which farmers received headage payments, a subsidy premium for every animal produced. This really turned on the taps, and as serious sums of money started to flow into the uplands, farmers did what any economically rational person would do: they increased their livestock numbers. Because sheep are smaller and faster growing, their numbers could be increased more easily than cattle, so sheep became even more dominant. For the two decades that the scheme operated, the headcount skyrocketed, with the number of sheep and lambs in Cumbria peaking at 2.6 million by the mid-1990s, well over double the average at the start of the century.

These subsidies were well intentioned, but badly flawed. The costs of producing livestock in the uplands are higher than in the lowlands for a whole range of reasons. Grass and other plants grow more slowly and are generally of lower nutritional value, so livestock are spread thinly over large areas, which means high labour costs for even basic animal husbandry. In the great national drive for agricultural production, the hill farm's higher costs and lower outputs could never compete. The subsidies were intended to prevent these fragile farms from going bust, and to protect the social fabric of rural upland communities of which the farms were such an important part.

Production-based subsidies changed the logic of hill farming. The most profitable farms were those that carried the largest numbers of animals and claimed the biggest payments. Many hill farmers suddenly had the financial freedom to farm in a very different way. They now had the resources to buy in feed from elsewhere and to increase yields of their own crops using synthetic fertilizers. New medicines and nutritional supplements helped to boost growth and ward off disease. Quad bikes, tractors and powerful 4x4s started appearing everywhere, massively reducing the time and effort involved in checking on the stock, taking them to market, cutting, baling and moving the hay. As the operations became slicker and as the productivity of the land was incrementally increased, it made sense to carry on raising the livestock numbers. As the flocks grew larger, so did the subsidy income, which could be spent on boosting production still further, breaching the thresholds that had kept nature in balance with farming for generations.

What didn't keep up with the scale of the expansion was the price of the produce. For thousands of years, sheep had been favoured because, unlike other animals, they yielded both food and fibre. A lamb and a fleece every year make for quick returns compared with a slow maturing cow. It would be hard to over-estimate the economic and cultural value of wool to British society. The wool trade was of vital importance to our national prosperity. Wars were fought over it and wealthy landowners counted their riches in sheep numbers, rather than money. Without wool and the role it played in the industrial revolution, places like Leeds and Bradford would not be what they are today – they are cities built from wool. The invention of cheaper, easier-to-produce synthetic alternatives means wool is no longer as highly valued – it typically costs more to shear the sheep than the farmer earns from selling their fleeces. This is a tragedy. Wool is a fantastically durable, natural material, but supply has for decades been outstripping demand, making it close to worthless for many sheep farmers. For most businesses, a drop in demand usually means a drop in production, but subsidies decoupled these basic market forces. Once costs of production are considered, lamb and mutton are often similarly unprof-itable, but with businesses propped up by government support, none of this mattered, and the output carried on growing.

As sheep numbers rose, the workload inevitably increased too. Farmers are without doubt the most hard-working people

I know, typically at work from dawn to dusk and often beyond. I'm constantly ribbed by Matthew for keeping regular office hours, a friendly slur slung into the window as I drive out of the yard at half five. A farmer's graft is driven by a love of the job, but in recent decades it also became a necessity of the relentless system created by headage payments.

Bill was working at Haweswater when sheep numbers were at their peak. Naddle Farm carried in excess of 3,000 ewes at one point. In order to get around the fact that land was limited within the farm boundary, there was a much greater reliance on the commons to accommodate the sheep year-round. This left the flock badly exposed when severe weather hit. Flying feed blocks up onto the fell by helicopter was a regular occurrence, to compensate for the poor nutrition in the skeletal winter vegetation. The livestock losses must have been enormous, but with government money continuing to pour in, it wouldn't have made any real difference to the business. The abundance of dead sheep on the hill likely kept the Haweswater eagles well fed during this period. There's a complicated irony here – sheep were preventing the recovery of habitats that would have supported wild prey for eagles, but they were probably also what was keeping them alive.

An incredible amount of harm was inflicted during this time, but as is so often the case, nobody realized until it was too late. All of a sudden, it seemed, the lapwings, grey partridges, curlews, corn buntings and corncrakes – birds that had been the companions of farmers since time out of mind – weren't as common as they used to be. The skies were emptying, the countryside being muted. During the last three decades of the twentieth century, subsidies did more damage to the ecology of the uplands than had occurred since Neolithic times.

The sheer grind of endlessly increasing output took its toll and some farmers couldn't keep up. The ancient patchwork of small farms declined, with the survivors subsuming neighbouring land and selling off surplus farmhouses into the developing property market. Rather than protect the fabric of rural upland communities, subsidies accelerated its breakdown, a process that's still playing itself out.

Because of the lucrative earning potential offered through grants, land, already considered a safe bet, became an even more attractive investment opportunity, driving up land values, and further concentrating ownership in the hands of the few. This is reflected nationally: the UK's farmland is among the most expensive in the world and has more than trebled in value over the last twenty years. Combine this with the fact that the UK has the largest average farm size in Europe, and it becomes easy to understand why it's so difficult for new entrants to farming to get going. What young farmer could afford to buy or even rent the supersized farms that now dominate? So, the stock of remaining farmers grows older – their average age in the Lake District today is fifty-nine.

I'd love to own land to nurture and tend. I only dream of a small patch, somewhere Becki, the kids, a scattering of livestock and I could sink into. We'd farm like people did a century ago, without chemicals or fuel-hungry machinery. We'd strive to make our land scruffier, woodier and wetter, in the hope of sharing it with as many flowers and wild creatures as possible. But without a change in land prices or some impossible windfall, a dream is all our farm is ever likely to be.

The impact of millions of mouths nibbling away at fellside pastures would have been hard to detect initially. Land wasn't completely stripped, a green cover remained throughout, and from a distance the view of the landscape in 1950 would have been hard to distinguish from the one in 1995, when sheep numbers reached their peak. Up close, though, things had definitely changed. Hillsides once full of flowers and their attendant insects became overrun with moor mat-grass, purple moor-grass and heath rush. The contrast is plain to see at Haweswater wherever there is a fence line that separates areas with different levels of grazing. One broad section of verge approaching the bridge across Rowantreethwaite Gill is full of heather, bilberry, wood sage and bedstraw; the other side is tough grass and bracken.

These coarse grasses aren't just rubbish for wildlife. Generally speaking, they make poor grazing too – the sheep avoid them for good reason. The reduction in the quality of the sward caused by the inflated numbers of sheep not only stole from nature, robbing the pollinators of their flowers and the birds of their seed and insect food, but they were also eating away at the health of future generations of livestock – the generations living now.

As the diversity in these pastures took a hammering, their physical structure also changed. With lower numbers of animals on the hill, there would always have been variability in the growth of the plants. Close-cropped areas where the

Bilberry

vegetation was particularly appealing would have existed in a mosaic alongside tangles of dwarf shrubs, stands of tall grasses and scrubby patches, providing nest sites, food and shelter for stonechats, linnets, reed buntings, cuckoos, grouse and hen harriers. The lawn-mowing effect of the sheep converted thousands of hectares to close-cropped baize, suitable only for meadow pipit and skylark. When we carry out our upland bird surveys on the fells around Haweswater, we can walk for miles and record only these two species. I've got nothing against meadow pipits or skylarks – their joyful songs and springtime squabbling are montane delights – but when they are the only birds you see, you know that you're in a habitat that's in poor health.

❧

Monitoring change is a big part of our work. Across the three upland commons that we hold grazing rights for, we've established eleven 1km-square survey plots. We walk fixed transects within these plots twice each year, recording all the birds that we see.

One of these surveys stands out in my memory as the moment when the paucity of birdlife in our uplands was really driven home. I was covering the two plots that sat adjacent to each other on Bampton Common, bound on their western edge by the remains of the Roman road running along High Street. There were six miles of north-south transects to be walked, across some of the highest land in the area, from Randale Beck, which drains the valley below Kidsty Pike, up over the ridge between High Raise and Low Raise to Longgrain Beck, which drains the next valley to the north.

Fine days with good visibility are needed for these surveys

to ensure the best views of the birds, which tend not to be as active in cold or wet conditions. Hiking in the mountains in decent weather looking for birds may not sound like hard work, and I'm certainly not complaining. The setting was unquestionably stunning. I strode up and down the sides of steep grassy valleys, rocky streams winding along their floor. From the flat summits, the central Lakeland fells to the west were like giants' molars chomping at the sky.

It was an hour's walk to the start of the first transect from Mardale Head, and I'd set off at first light to give me time to complete the survey during the morning when the birds would be most vocal. Following prescribed transects goes against normal walking instincts. I was forced to follow routes that led straight up and down slopes, rather than picking out gentler zig-zag paths as I would have done otherwise.

With eyes and ears open, I made steady progress, glancing at my handheld GPS to ensure I was still on the transect, stopping regularly to scan with binoculars and to mark what I'd seen or heard on the clipboard-mounted map slung around my neck. Meadow pipits and skylarks were as ubiquitous as ever, their nests hidden in the grass and their songs my perpetual soundtrack. I'd recorded several pairs of sharply dressed wheatear, mostly around boulders alongside Randale Beck, under which they would have secreted their pale blue eggs. Appropriately enough, I'd seen a big group of ravens on the flat summit of Raven Howe. These would have been a party comprising a few family groups, including that year's young, which stay with their parents for six months after fledging. The collective noun for ravens is an unkindness, which says a lot about our attitudes towards these intelligent birds. A dipper flew over me as I sat for a breather at the ruined sheepfolds next to Longgrain Beck.

Having walked this route annually for the last few years, I didn't have high hopes for anything new, and I got what I expected. Two carrion crows near the end of the final transect completed my tally. After five hours of walking and searching, I'd recorded six species.

&

The day before I'd done my survey, Spike, one of the wardens at Haweswater, led a guided walk in Naddle Forest for International Dawn Chorus Day, an annual event that takes place on the first Sunday of May. Spike's ornithological knowledge is immense, and his ability to identify birds by the most fleeting fragment of song is extraordinary.

I found Spike a little intimidating when I started at Haweswater. Dreadlocked and weathered by years of upland outdoor working, there's something flinty and implacable about him. A skilled drystone waller, grower of juniper and planter of trees, Spike is a formidable part of our little team, with a huge capacity for work in even the grimmest of weathers.

Spike has been surveying the woodland birds at Haweswater since 2006, a job he took over from the previous warden, making this one of the longest-running studies of its kind in the country. Each spring, Spike visits a series of predetermined points scattered throughout the woodland and records the birds that he hears, placing them in different distance bands from his location. Because birds are so hard to see in thick woodland, identifying them by their calls is the only workable option. It's skilled and important work. Being so in tune with the forest, Spike was the perfect leader for the dawn chorus walk, helping the attendees to understand what it was they were hearing. I was there mainly to bring up the rear and make sure we didn't lose anyone, as well as to provide an extra pair of eyes and ears.

Atlantic oakwoods like those at Naddle are incredibly special places. The Lake District's proximity to the Atlantic, and the resulting mild, wet climate means that we are one of a handful of places where temperate rainforest can be found, a globally rare habitat. The rich growth of mosses, ferns and lichens, which festoon the trees in Naddle Forest, are all rainforest indicators. As you'd expect in a rainforest, the bird life is rich too. Each summer, the resident birds are joined by migrants – including vibrant redstarts and dapper, two-toned pied flycatchers – visiting from Africa to breed in holes in the old trees.

We struck out from Naddle farmyard at 6 a.m. on a leisurely loop, taking in the places Spike knew would give the group the best chance of seeing or hearing most of the woodland's specialities. The real trick was picking out individual birds amid the clamour, which that morning was every bit as raucous as we'd hoped it would be.

Everyone could pick out the cuckoos, but Spike was on hand to help the group distinguish the garden warblers from the blackcaps, and to tune into the subtle calls of treecreepers and spotted flycatchers. We heard buzzards and blackbirds, wood and willow warblers, chiffchaffs and chaffinches, and caught a glimpse of a tawny owl late for its bed. After a charmed hour and a half, with the freshness of the morning still in our noses, we'd clocked up thirty-four bird species, as well as several red squirrels.

Two experiences on two consecutive days. A short morning walk through woodland where the birds were everywhere, in spectacular diversity and abundance, then a six-mile trudge where in comparison the place was almost birdless. You might say that I'm comparing apples with oranges. Of course there are more birds in the woods – the fells could never be as rich. This may be true, but the contrast needn't be quite so pronounced. Only 12 per cent of the Lake District is wooded, and half of that is non-native conifer plantation. Much of the sheep-grazed fells could be woodland, or wood pasture, or have pockets of scrub breaking up the grassy monotony. Shouldn't the multitudes in the woods have a little more space to sing?

Like huge swathes of our national parks, the land I marched across in search of birds on Bampton Common is access land, where everyone has a right to walk. This is a modern freedom, enjoyed by millions. Before the Second World War, we were far more constrained. Vast areas of Britain's countryside were off limits to the public, reserved by private estates or the Crown for shooting or farming. The mass trespass of Kinder Scout in 1932, where hundreds of ramblers marched across private land owned by the Duke of Devonshire, was a powerful expression of the public's dissatisfaction with the state of things.

Disgruntlement simmered during the war years, but in 1949 the National Parks and Access to the Countryside Act was passed with all-party support, as part of the post-war reconstruction led by the Labour government of the day. The act was the cornerstone of a completely new approach to nature and the countryside. It gave birth to the national parks, created the government's nature conservation agencies, and paved the way towards much of the environmental legislation that we have today.

That the Second World War resulted in the stimulation of two seemingly opposing forces – intensive farming and nature conservation – is a fascinating paradox. The tussle between the two is what has made caring for the countryside so difficult over the past decades, and it's only recently that strides towards some semblance of balance have been made.

The awareness of how much harm intensive farming was doing took time to develop, but by the middle of the 1980s plans to repair the damage were starting to take shape. The first agri-environment schemes came into being at this time. Initially only available in particular areas, they were gradually scaled up so that farmers anywhere in the country could apply. The principle was a simple one: if a farmer agreed to farm in a less intensive and more environmentally friendly way, the government would compensate them for the loss of agricultural income that resulted. These schemes have had multiple iterations and names over the decades since their introduction, but the principle of payment for income forgone has remained at their heart.

Current schemes last for five or ten years, and for many farms they can contribute a significant proportion of the overall farm business income. At Haweswater, we are involved with four of these agreements, one for each of our commons, and a fourth that covers our two farms, Naddle and Swindale. Because our focus is more on environmental protection than on production, these schemes provide the majority of our income – in a typical year they are worth more than four times the money that we get through the sale of livestock. A big part of my job is making sure that what we do at Haweswater gives a good return on the investment that the government is making in our land through these schemes.

The main way that agri-environment schemes can effect change on mountain and moorland is by placing limits on grazing. Sheep numbers started to drop in Cumbria in the mid-1990s. There was a sharper fall resulting from the foot-and-mouth crisis around 2000, but the population has since levelled off at around two million sheep, still almost double the pre-war population.

Government funding subsidized a massive assault on the countryside and its wildlife in the post-war decades. It's now being used to repair the damage that was inflicted. This all amounts to a complete about-face in government policy within a lifetime. If only we'd had a little more foresight, we could have saved ourselves an awful lot of bother.

I know local digger drivers who have blocked drains that they can remember digging decades previously, both operations funded by the taxpayer. This is a deeply unsettling position for farmers to be in. Imagine knowing that a change in policy or in government could mean that everything you've worked so hard for could shift beneath you.

❧

There are a growing number of prominent Lake District farming voices on social media, a handful of whom also write books and newspaper articles. I've met some of them in recent years, but you only have to read what they write for an insight into how threatened they feel their way of life is. Take for example a *Guardian* article from 2015 by Andrea Meanwell, a response to accusations that farming had 'sheepwrecked' the uplands. Andrea tells us how her family have been farming in Cumbria for 500 years, that the landscape has been created by farmers and sheep, and about all the hard work involved. Pride in what she does comes through strongly, as does her hurt at the criticism of her way of life.

I first met Andrea during an interview for BBC Radio Cumbria. We'd been picked as two people at either ends of the supposed 'rewilding vs farming' debate. While the interviewer set up the recording equipment in Andrea's cosy kitchen, we chatted over a cup of tea. It turned out that Andrea was due to be coming over to Naddle Farm the week after the interview for a lichen identification course that we were hosting. This led comfortably into amiable chatter about wildlife, sightings of otters on her beck, the restoration of her hay meadows and the comparable work that we were doing at Haweswater.

I suspect the interviewer was hoping that Andrea and I would lock horns. He kept trying to nudge the discussion towards the more inflammatory aspects of rewilding, like the reintroduction of carnivores. Since this wasn't a subject that had much influence on either of our day-to-day lives, we skipped over it. We agreed about much more than we disagreed, and all in all it was an enjoyable hour-long chat, much to the disappointment of the man with the microphone.

It was an odd way to get to know someone for the first time, but we've bumped into each other several times since through Andrea's job as a farm advisor for the Lake District National Park Authority. Like many other farmers in the uplands, she must have many strings to her bow to keep her farm going. As well as her work for the Authority, Andrea writes books and articles, knits woollen berets, keeps rare-breed sheep and earns funding for environmental work on her farm.

Farmer bashing is fashionable at the moment and Andrea's *Guardian* article was a very understandable response to it. It's all too easy to cross the line between blaming farming policy, which is the real cause of environmental damage, and blaming farmers who have simply been responding to those policy demands. If you don't know any farmers, it's easy to fail to appreciate how hurtful not making this distinction might be, and how counterproductive.

In my first years at Haweswater, most of my engagement with farmers had been in formal meetings, where a group mentality reigned. The combative nature of these meetings had made me defensive and unwilling to take the time to get to know my farming neighbours one to one. Of course farmers aren't all the same, just as all conservationists aren't. Nobody deserves to have their community vilified in the way that farmers often do, and in the way that social media makes so easy. But I can also understand where the frustration comes from. In the face of rapidly dwindling wildlife, millions of us have felt totally powerless. Pointing the finger of blame feels like doing something – we can tell ourselves that it's helping to raise awareness of the problems. This doesn't make it right, and, too often, it's more likely to fan the flames of a debate that's already overheated rather than drive any meaningful change.

The anger directed at farmers often appears to be coming from people living distant lives, the urbanites who make the decisions, who know how to organize, influence and campaign. Through sheer weight of numbers, they can drive changes to agricultural policy and affect the payments that keep farms running. No surprise that farmers like Andrea feel ganged up on.

Andrea's doing great things for nature on her farm. She's fenced out watercourses and copses, planted trees, and is restoring flower-rich banks and meadows. It isn't fair that the simplistic critique of upland farming ignores her efforts.

As the evidence for the link between intensive farming and wildlife decline became incontrovertible, rewilding as a means to repair some of the damage grew in popularity. Rewilding as a concept has been around since the 1990s, but it was given a significant boost with the publication of George Monbiot's book *Feral* in 2013.

Feral makes compelling reading, powerfully arguing for a reduction in human intervention to give nature the freedom to look after herself. That this blindingly obvious concept even gets discussed must seem peculiar to residents of nations that still contain areas of genuine wilderness, and it's perhaps a sad reflection of the state of our islands that rewilding has become so controversial.

The growing and genuine concern about the decline of nature and the powerful optimism of rewilding was a potent combination, and the concept quickly gained a huge amount of traction. George simultaneously became the unofficial champion of a reinvigorated environmental movement and public enemy number one for many farmers.

As with so much in modern life, *Feral* was turned into a soundbite. George's words were twisted so that many believed he was calling for rewilding to replace farming everywhere. This isn't what *Feral* says. Rewilding Britain, the organization that George helped to establish to try to bring the book's principles to fruition, wants to see 5 per cent of the country become core rewilding areas, and 30 per cent of our land and sea to be restored for nature by 2030. That doesn't sound so threatening to me, especially when you consider that there is no mechanism for forcing this change on anyone. Rewilding will only happen where landowners and managers want it to happen – no one is going to turn up at your door with a bunch of willow cuttings and a gun.

It would be pointless to deny that I'm fascinated by rewilding. In the right places, I believe it offers incredible opportunities to restore wildlife and habitats at scale. But I loathe the toxic polarization that rewilding has fuelled. There doesn't have to be a fight between rewilding and farming. I'm hoping that our work at Haweswater can show that it's possible, indeed preferable, for high-quality livestock production to sit alongside land managed with a very light touch, where natural processes operate and wildlife thrives. You can decide for yourself if you think this qualifies as rewilding or not.

⁂

My interest in rewilding goes back a long way. In 2005, Becki and I spent several months travelling around the Highlands of Scotland in a van with a bed in, researching people's attitudes towards mammal reintroductions for my master's dissertation. I had been living and working in London for a couple of years after completing my undergraduate degree, working in a warehouse and moonlighting as a musician. The MSc was a bid to get my career working with nature back on track.

Reading about reintroductions that had been happening with increasing frequency across the globe, and the growing calls for the UK to follow suit, had sparked my interest. I'd heard about an organization called Trees for Life which had developed a compelling 250-year vision for ecological restoration in the Scottish Highlands. The area they had reimagined included some of the wildest land in the UK, centred on Glen Affric and its remnants of native pinewood. Stretching from Inverness in the east to the Isle of Skye in the west, this is one of Scotland's most sparsely populated areas, with minimal road infrastructure, and hundreds of square miles of moorland and mountain. Trees for Life's vision saw native habitats restored at a grand scale, and the return of many of our extinct native species, including beavers, wild boar, lynx and wolves.

I wanted to gauge levels of support for the return of these

missing species in a place where there was a chance of their reintroduction occurring at some point in the future. Using a standardized questionnaire, I interviewed residents of properties selected by randomly generated grid references, which made for an interesting way to see the countryside. I inadvertently visited plenty of ruins and barns, not being able to distinguish between abandoned and occupied properties on the map. I was more than once seen off by a pack of farm dogs, and on one occasion was followed back to the van from the door of a baronial hunting lodge by a herd of semi-tame red deer hoping for a feed. I spoke to some people who were convinced that lynx were already roaming wild, and others so appalled by the mere suggestion of reintroductions that they defaced my questionnaire with angry expletives. But it all worked out, and the completed questionnaires allowed me to categorize respondents into recognized 'attitude typologies', which described how they thought and felt about nature.

The farmers and gamekeepers who completed my questionnaire had a different set of attitudes to the rest of the respondents. They tended to conform to the humanistic, dominionistic or utilitarian typology, meaning that they viewed the natural environment from a human point of view, felt dominant over it, and saw it as a resource to be used for human benefit. Meanwhile, the general non-farming rural resident conformed more to either the naturalistic, ecologistic or aesthetic typologies, meaning that they saw the natural environment as important for its own good, recognized the value of the connections between living things and saw nature as a source of beauty and inspiration.

Like much in science, this research did a good job of stating the blindingly obvious. It's no great revelation that farmers see the environment as something to control in order to provide for the human race, while much of the general population see it in more benign, passive terms. The research concluded that there was broad support for species reintroductions in the general populace in the study area, but that farmers and gamekeepers were more opposed. Despite being few in number, the latter could have a huge influence over the success or failure of species reintroductions.

This range of intrinsic attitudes is part of what makes discussions about caring for our land at Haweswater so difficult, so rancorous. To understand these differences is not to widen the split between farming and conservation camps, it's the opposite. My hope is that increasing respect for different points of view will lead to an acceptance that the only way forward is for sustainable farming, rewilding and all the approaches in between to coexist. If we are to respond to the joint climate and biodiversity crises we find ourselves in, farmers, landowners, government, businesses, charities and anyone else who cares must work shoulder to shoulder.

No one person can bring that perfect world into being. It will be made by countless individual actors and their innumerable decisions about how to tend to their own land, nudged by policy and people, economy and ecology. I'm one among many, and there's still much to learn from the past and the present to determine what part Haweswater might play in that sunlit future.

CHAPTER 5

The Corpse Road

ROWANTREETHWAITE:
The clearing with the rowan tree
(Old Norse/Modern English)

A series of ruined stone huts sits on a fellside shoulder of Mardale Common, 100 metres above Haweswater's surface. The most intact of them lacks only a roof, its substantial stonework standing proud, seventy years after timbers that supported heavy slates gave way. It has two small square windows, and a huge stone lintel above its low, empty doorway. Sheep make use of it for shelter from the wind, and their dung fertilizes a tight mat of nibbled grass and rush tussocks growing out of the packed earth floor. A little to the north is a second building, this one with two rooms. It hasn't fared quite so well, and the walls are starting to crumble around its door openings, now that the lintels have gone. If you search in the bracken that grows thickly on the slopes surrounding the buildings, older remains are hidden. Most are little more than rectangles of low rock heaps, footprints of structures that probably had their stones recycled into one of the buildings that still stands.

My son Elliot and I enjoyed a lunch of pot noodles and a flapjack in one of these buildings a couple of Decembers ago, ancient walls providing a handy windbreak for the camping stove. It wasn't the most well-planned of meals, the result of a hurried raid on the kitchen cupboard and a supplementary stop at the petrol station in Penrith, but it filled a hole.

Naturalists like me don't switch off during leisure time. Our eyes are always scanning for birds in the sky, for flowers underfoot, for clues as to what made a habitat the way that it is. I learn as much about nature on walks with my family as I do when I'm out surveying. Sharing our observations to enthuse each other doubles the pleasure of it all.

We'd taken an adventurous route to reach our picnic spot, via the primordial gorge of Rowantreethwaite Beck, which drains a boggy expanse of Mardale Common and flows into the reservoir's steep eastern shoreline. Tumbling over waterfalls and mossy boulders, through secret pools, bifurcations and reconnections, Rowantreethwaite Beck, together with the deep cleft into the landscape in which it runs, creates a rich and vibrant fragment of wildness.

In the summer, the sides of the gorge are thick with ferns and wildflowers, and a tangle of branches reach back and forth over the rushing water. One of our largest surviving wych elm trees grows out of a rock face, and a colossal yew casts deep shade a little further up. Dippers zip up and down over banks studded with early purple orchids, primroses and common bugle. Beech fern, hard shield fern, black and maidenhair spleenwort protrude from cracks in the rock, which are blotchy with crusts of multicoloured lichens.

On a survey of the gorge in 2019 with Rob, a gifted young local botanist, we recorded close to 200 species of flowering plants and ferns. Its richness is attributable to its steepness – few sheep or deer are prepared to risk getting down the valley's sheer sides, and so the plant life grows unmolested.

In a sheltered dell a few hundred yards up the gorge, where a carpet of wild thyme, wood sage, hawkweeds and foxgloves was growing over a clutter of scree, I had a moment of serendipity. Rob was out of sight, surveying the moorland above, and so I entered the dell alone, just as the afternoon sun slanted in, illuminating flowers at their midsummer peak, bursting with nectar and at their most attractive to insects. Energized by warmth and sugar, a vast cloud of butterflies had descended, zooming from flower to flower as if their short lives depended on it, which I suppose they did. Painted ladies made up the bulk of the gathering, having arrived in the UK that year in great quantity, advancing north on their long migration from Africa. They argued over access to the flowers with red admirals, small pearl-bordered and dark green fritillaries, large and small whites, small heath, common blue, as well as silver Y and brown silver-line moths. The chirruping of grasshoppers and the drone of hoverflies and bumblebees mingled with the Mediterranean scent of the thyme. The kaleidoscopic blur of insect wings over the flaming purple and yellow flowers was hypnotic. I sat and gawped, trying with little success to capture the moment with my camera, while the painted ladies tried to pollinate my T-shirt. Then the sun sank below the lip of the gorge and following the light, the insects flew up and out of view. The party I'd stumbled into was over. None of these species that had me entranced are particularly rare, but their abundance was breathtaking, concentrated into this sunlit pocket of nectar-rich habitat.

During my trip with Elliot in December, we didn't have flowers or butterflies to distract us, but Rowantreethwaite was still enticing. The gorge isn't wide – on an aerial photo it looks like little more than a single row of trees – but once inside, it envelops you completely. I'd told Elliot that the gorge's primeval tangle of trees and rocks might hide dinosaurs. This was more than enough to keep his seven-year-old imagination firing as we scrambled upwards and onwards, sometimes hand over hand, keeping our eyes peeled for archaeopteryx nests and velociraptor prints.

Eventually we emerged into open ground. Having been cocooned in the gorge, our view limited by its tree- and fern-clad sides, suddenly being able to see for miles in every direction felt vaguely agoraphobic. Across the reservoir, we had an uninterrupted view into Riggindale, the prominent peak of Kidsty Pike overlooking its north-western corner. Bampton Common extended away to its north with a palette of grey rock, fading yellow-green grass and blotches of ginger bracken.

A breeze had picked up while we'd been concealed below, so the shelter of the stone huts beckoned. The fells around Haweswater are peppered with ruined stone buildings. They are part of what gives the landscape its air of lonely, wild drama. Many of them are thought to be 'peat houses'. Woodland, which clothed the prehistoric fells, had been largely cleared by the end of the Bronze Age, only clinging on in steep and inaccessible places where felling the trees wasn't worth the risks of reaching them. So, for most of the past 3,000 years, in the absence of readily available timber, peat has been an important source of fuel.

Peat is the semi-decomposed remains of mosses and other plants that forms in wet and acidic conditions. It is much harder won than timber, requiring a huge effort to cut, cart and dry, but its high carbon content makes it an effective fuel nonetheless. On being cut from the bog, the peat is sodden, heavy and difficult to move, so fellside peat houses were often built where it could be stored for drying. The dried bricks of peat would later be taken down the hill on the backs of fell ponies.

Peat cutting isn't practised in the Lake District any more, the ease and efficiency of fossil fuels and electricity having made it redundant. However, once you get your eye in, you can see evidence for just how widespread the practice used to be. Nature tends not to work in straight lines, so a linear bank in a bog is almost certainly the remains of a peat working. Peat was not just for heating people's homes and cooking; it also provided the fuel for smelting lead, copper and tin, for pottery and lime kilns, and for the production of tiles and bricks. The scale of the extraction was huge. A reliable account of copper smelting near Coniston states that 4,800 horse loads of peat were used in a single twenty-week session.

Although it was an unmechanized process, over the course of centuries unimaginable volumes of peat were stripped from the landscape, its combustion sending huge quantities of carbon dioxide into the atmosphere. Peat being so rich in carbon, burning it is every bit as bad for the climate as burning fossil fuels. Where bogs haven't been drained, the layering of vegetation means that peat can regrow, but only at about one millimetre per year. That's a lot faster than the formation of coal, oil or gas, which takes millions of years, but peat can hardly be described as renewable.

Elliot's name for a bog is 'splodgey-pond'. It probably won't catch on, but it nicely captures how this habitat is neither quite liquid nor solid. Beneath our feet, as we splashed from the top of the gorge across to the huts, was a fascinating and under-appreciated array of highly specialized plant and animal life. The acidic and hyper-saturated conditions in bogs mean that they contain very few nutrients, so most of the plants that grow here are small and inconspicuous. Many species are adapted to life in the bog and can survive nowhere else. One select group of plants has evolved a novel way to supplement the lack of goodness they can obtain via their roots: they eat insects.

Sundews have vivid red tentacles protruding from the ends of their paddle-shaped leaves, out of which they secrete a drop of sugary fluid resembling dew, which lures in tiny insects. Little do the bugs realize that these 'dew' drops are filled with a sticky, glue-like substance, strong enough to stop midges and other tiny victims from escaping. With agonizing slowness, the leaves fold over the captives, bringing them into proximity with specialized glands that dissolve and digest them. Charles Darwin did much to shine a light on the carnivorous habits of these hitherto little-known botanical monsters. He became obsessed with them, completing countless experiments and drawings, stating at one point that he cared about them more

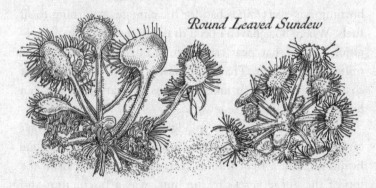

Round Leaved Sundew

'than the origin of all the species in the world'. Elliot saw their appeal too, prodding at their sticky leaves to see if they might fancy a nibble of his finger.

Butterworts do something similar, with their sticky secretion coating their thick green leaves, which form a ground-hugging star. Butterwort's striking, violet-like flowers are held well above the leaves to avoid inadvertently consuming their pollinators.

Even more impressive and surprising are the bladderworts, of which there are several species in the UK. Bladderworts live in open water, so have dispensed with roots altogether, floating free in boggy pools. Along their rootlike, yellow stems, thread-like leaves are interspersed with tiny hollow spheres. By moving water from the inside to the outside of these bladders, they become filled with air, helping the plant to float, but also enabling it to perform a deadly trick. The bladders are fitted with a tiny trapdoor, surrounded by a ring of sensitive hairs. If some aquatic insect or tiny worm brushes past them, the trap-door opens, and because of the lower pressure inside the bladder, the hapless creature is sucked in and the trapdoor slams behind it. The whole process happens with astonishing speed, in less than 0.03 of a second. Enzymes in the bladder dissolve and consume the trapped insect. In less than thirty minutes, water is pumped back out of the bladder and the trap is reset.

Many groups of more familiar flowers have a member that is a bog specialist. At the head of Swindale there is a thriving colony of bog orchid. Rarely growing longer than the length of your thumb, they are Europe's smallest orchid species. Finding them takes serious dedication and a willingness to get wet knees, as the only way to pick out their yellow-green flower against the background of other plants is to get down on their level. Then there is bog rosemary, bog pimpernel, bog stitchwort, bog bilberry, bog pondweed, bogbean, bog myrtle – you get the idea.

Cottongrass is one of the few bog species that is hard to ignore. Its fluffy white seed heads fly like flags over the bog in the summer months and provide a source of soft nest-lining material for moorland birds. They last for months – a tiny bouquet of them picked by Elliot from our local bog in spring lasted well into winter.

The most important species that grow in peat bogs are the bog mosses, or sphagnum mosses to give them their proper name. As you walk across a healthy bog, it will be sphagnum that makes up the bulk of the life under your feet. There are thirty-four different species in the UK, each with slightly different demands and habits. Most haven't been given common names, which along with the fact that they are small and unassuming and grow in places where you stand a good chance of over-topping your wellies, means they don't have a high profile with the general public. Yet they have a subtle beauty and a significance that far exceeds their size. I've been on several sphagnum identification courses, but only a handful of species ever seem to stick in my cluttered memory. One is the vivid green *Sphagnum cuspidatum*, the species that most readily colonizes open water and according to my instructors is supposed to have the appearance of 'drowned kittens'. The most impressive is *Sphagnum magellanicum*, which on intact bogs forms deep-red hummocks.

Sphagnum has a spectacular ability to hold onto water. Grab a handful, and even after a dry spell you'll likely be able to squeeze a good quantity of water out of it. Dried sphagnum is able to absorb twenty times its own volume of water, and it's been used through the ages as a wound dressing and for feminine hygiene. The Romans kept it next to the loo. With sphagnum being its primary constituent, peat also has incredible water-holding capacity, which is one of the reasons why it's so popular with gardeners. An upland landscape with lots of peat and sphagnum therefore has an unparalleled ability to regulate water as it flows down the hill.

The importance of sphagnum and the peat that it forms can't really be overstated. There is more than ten times as much carbon stored in the peat soil of the UK than in all of our forests put together. And yet, across the globe, the UK included, peat soils are in very poor shape. For generations we've been cutting and burning peat, draining it for agriculture, harvesting it for gardening and setting fire to it, accidentally or otherwise. Damaged peatlands are a massive contributor of global emissions. If they're dried and exposed, peat soils release carbon faster than they are able to capture it from the air. When they're wet, the opposite is true. With all that we know about climate breakdown, and about how much of an ally peat can be in fighting it, it is astounding that harvesting peat for horticulture continues to this day and that you can walk into any garden centre in the country and buy a bag of the stuff. A ban on the extraction and sale of peat can't come quickly enough.

As you trudge through an intact bog, the variety of sphagnum forms a multicoloured undulation of hummocks and hollows, over which fragile tendrils of cranberry creep. Snipe, skulking brown wading birds with a fondness for bogs, often burst out from under your feet, seemingly waiting for the last possible moment to shock you. Dragonflies breed in pools with emergent bogbean, which give shelter to frogs and toads, while common lizards and adders bask at the drier edges. If you jump in the right places, the whole bog can shudder, reminding you that the metres of peat below are only tenuously solid.

An intact bog helps to regulate the climate, contributes to reducing flooding and provides a home for an incredible array of wildlife. A bog that has been drained or had its peat cut away does none of these things. Hindsight makes it easy to criticize historical practices, but of course the Lakelanders of old had no way of knowing that burning peat might have an effect on the environment. Even if they had done, they had little choice but to do it – cutting peat was a matter of survival.

The huts on Mardale Common weren't peat houses, though, or at least, that wasn't their sole function. Generally speaking, peat houses don't comprise more than one room, or have windows. The presence of these features suggests that these huts were for people to sleep in. Today, they make a handy place for a picnic on a windy day.

❧

The zig-zag footpath that ascends from the reservoir up to the huts is an ancient route, cheerfully known as the Old Corpse Road. Although the now flooded hamlet in Mardale had a church, it didn't get a consecrated burial ground until 1736. Up until that point the valley's people and their ponies had to convey their dead up over Mardale Common, into Swindale and then onward to the nearest cemetery at Shap. The route is now a public right of way, passing between the stone huts. They make a popular photo opportunity, the melancholy of their remains set against the grandeur of the mountains beyond.

Even with the help of a fell pony, carting a cadaver up over terrain as steep and rugged as this would have been a slog. But the huts are less than a mile from where the church used to be, so it seems unlikely that the cortege would have needed a place to sleep so soon after setting out.

I have my own theory about the function of these buildings, based on their name – High and Low Loup. The place names of Cumbria are a fascinating muddle, reflecting waves of settlement and occupation by different cultures through time. There are places with Norse roots, reflecting the influence of the Vikings, and others with Old English, Celtic, Brittonic, Anglo-Norman and Middle English origins. There are also many names that hark back to the Roman period. High Street, originally Via Alta, is the most prominent of these at Haweswater, which Elliot and I could see from Low Loup on the western horizon.

In Scotland, places with *loup* or *loop* in the name often refer to parts of rivers where fish leap. Even though the huts are only 100 metres or so from Rowantreethwaite Beck, it is so deeply incised into the land that it can barely be seen. Besides, the waterfalls below Low Loup are far too sheer for fish to jump, so this seems unlikely to be the source of the name.

There are many places in Cumbria that derive their names from animals, and my theory is that 'Loup' comes from the Latin for the Lake District's now extinct top predator, the wolf, *Canis lupus*. 'Loup' is still the name that the French use for the animal today. There are many wolf places in Cumbria, but these generally use the Norse root, *úlfr*, courtesy of the Vikings. Ulthwaite Rigg, a fell just south of Swindale, means the ridge with a clearing haunted by wolves. Uldale is the valley of the wolves, Ulpha, the wolf's hill, and Ullock, the place where the wolves play. Lake District places with Whelp in their name, such as Whelpside and Whelpo, refer to wolf pups. Clearly there were plenty of wolves in Cumbria when the Vikings started dishing out place names. High and Low Loup could have been given a Latin-derived name either by the Romans five hundred years before the Vikings showed up, or by the

Normans who arrived a couple of hundred years after the Vikings.

No one knows exactly when wolves disappeared from Cumbria, but they certainly persisted into the 1300s. A frequently cited piece of folklore states that the last wolf in Cumbria was killed in 1390 at Humphrey Head, one of the limestone peninsulas that juts out into Morecambe Bay in the south of the county. They held out in Scotland for much longer. Official reports state that the last wolf in Scotland was killed in 1680, but they were rumoured to have lingered on for a further two centuries. Whatever the exact timeline, there's no doubt that wolves and people once coexisted in Mardale, probably uncomfortably.

In popular culture, wolves are portrayed as slayers of grandmothers and little pigs, malevolent creatures of the wild, best kept from the door. In real life, their presence has a dramatic and positive influence on the habitats in which they live. Without top predators such as wolves, ecosystems simply don't function as they should: herbivore populations grow unchecked and the potential for natural regeneration of habitats is reduced. Wolves would have been the main predators of red and roe deer in Britain. In the absence of wolves, deer populations have exploded; their numbers are now higher than they've been for a thousand years.

Like most successful predators, wolves are opportunists, taking advantage of whatever prey they can find. In addition to deer, the wolves of Britain would have eaten rabbits, hares, foxes and badgers as well as smaller mammals, birds and amphibians in lean times. It's a safe bet that on occasion they would have also dined on mutton.

The ancient ancestor of the domestic sheep is the mouflon, a stocky white and tan beast sporting spectacular curly horns, more closely resembling a butch goat than a modern sheep. Mouflon rams exhibit dramatic combative behaviour, head clashing to demonstrate their dominance in order to win mating rights, much like rutting red deer do. Over the centuries, we've bred out much of this vim to create sheep that carry more meat, milk and thicker, more valuable fleeces, and which are less likely to break your ribs.

Modern domesticated sheep aren't defenceless by any means. The rams, or tups, regularly fight with each other, sometimes fatally. Groups of ewes still flock together, and squabble and headbutt in order to establish a social hierarchy, echoing the behaviour of their wild ancestors. Nevertheless, their relative docility, and the extra weight of flesh and fleece, means that compared to wild animals, domestic sheep make easy pickings for predators like wolves.

Having hampered their ability to adequately defend themselves, we had little choice but to become the sheep's protectors – and so the shepherd, one of the world's oldest professions, came into being. I can think of few occupations which have had a higher profile across human cultures than shepherds. Jesus was the Good Shepherd, the Prophet Muhammad worked as one as a young man and stated that all other prophets had done the same. The solitude and the simplicity of the life of a shepherd purportedly opened these holy messengers' minds to receive the word of God. The shepherd has become the embodiment of loving care, tending to frail and helpless creatures, at one with wild nature and free from the wicked influences of human society.

For millennia, shepherds have taken advantage of their woolly charges' innate fear of wolves by employing their domesticated kin – the sheep dog. Gathering flocks from the Lake District fells is an impossible task without their help. The unity of purpose between human and dog as the sheep are sent flowing down the hillsides is an undeniably impressive sight, and the skill of the dogs and their masters is formidable.

I regularly come across sheep in places where they're not supposed to be on our farms at Haweswater, in areas earmarked for regeneration of woodland or in the meadows while the hay crop is growing. Trying to get them back to where they should be without a dog necessitates much shouting, arm waving and mad dashing, and hammers home just how essential well-trained sheep dogs are.

Sheep dogs used to be more than just a tool for rounding up sheep. In regions of the world where large predators haven't been driven to extinction as they have in the UK, livestock guardian dogs are employed. Various breeds of dog are involved, all of which are large and powerful, with mastiff-like builds and protective tendencies. Puppies are introduced to the flock or herd that they are to protect at a young age. Once bonded, the dogs think of themselves as being part of the flock, and spend most of their lives in its midst. Long-haired and white, the Patou, or Pyrenean Mountain Dog used in France and Spain, even looks like one of its charges – very nearly a wolf in sheep's clothing. When a predator comes near, the dogs spring into action, barking aggressively to see the intruder off, or attacking them if the display of force isn't effective. In parts of Spain, Italy and Turkey some livestock guardian dogs wear wolf collars, fitted with vicious metal spikes designed to protect the dog if things get physical.

In many parts of Europe, wolves are on the rise as a result of increased protection and reintroduction efforts. This isn't universally welcomed, but attempts are being made to find a balance between the protection of farming livelihoods and ecological restoration. Some governments offer compensation for livestock lost to wolves, and a range of conservation organizations are offering training and support to reinvigorate husbandry that might reduce wolf-human conflict, like the use of livestock guardian dogs and active shepherding.

That the ultimate symbol of rewilding is driving a resurgence in traditional farming practices is an interesting paradox. This is doubtless one of many examples of how our ancestors managed to live more harmoniously alongside nature. Whether the farmers of the Lake District ever used dogs to protect their flocks from wolves is lost to the mists of time, but it's beyond doubt that more active shepherding was a much more prominent practice than it is today.

There are some well-preserved sheepfolds just to the south of Low Loup. A hundred feet by fifty, their stone walls are still mostly intact, indicating that they were in regular use until recently. The size of these pens suggests they might have been used to keep sheep in overnight, when the risk of predation would have been greatest. They sit in the middle of a wide, gently sloping bowl of land, probably the richest piece of pasture on the common, and one which offers no cover to an approaching predator.

Even without predators, shepherding played a vital role that I can't help but feel is an important missing component from modern hill farming. Actively shepherded flocks can be moved around to make use of the best areas of grazing and to avoid the worst. Mardale Common, the land surrounding High and Low Loup, is a landscape of boggy hollows and dry rocky knolls. There are steep cliffs and open water, fast-flowing streams, deep gorges and waterfalls, tick-infested bracken beds and patches of thick, spiny juniper scrub. The shepherds would have known the terrain intimately, its potential hazards and sweet spots. They would have kept the flock away from the cliffs and crags to avoid fatal tumbles, and avoided the bogs, where dark depths can be hidden beneath innocent-looking patches of moss. If a sheep did get itself mired, the shepherd would have been on hand to rescue it.

By keeping sheep on the move, shepherds ensured that there was always fresh, nutritious grazing to be had. Patches of pasture would have received pulses of grazing, followed by periods when the vegetation was free to grow, improving the quality of fodder, benefiting the soil and giving plants the chance to flower and set seed.

There's a fair chance that shepherds wouldn't just have had sheep for company. Cattle were a big feature of Lake District farming until relatively recently. You don't have to study an Ordnance Survey map of the Lakes for long before you see a place with a name relating to cattle. In our valley there are several, the most obvious being Butterwick, a hamlet on the road to Penrith. We have a field known as the Cow Pasture at the head of Swindale and many of our old stone barns are fitted out to house cattle.

There were goats too. The triangular summit of Kidsty Pike is *the peak at the top of the steep path where the young goats go*. I like goats, but I'm glad we don't have them roaming free in the Lake District like they do in some parts of Scotland and North Wales. Being that much nimbler than sheep, goats would put our fragile mountain flowers under even more pressure than they are already. Swindale means *the valley where the pigs graze*. The Lake District is littered with Goosemires and Goosegarths. The evidence for how much more diverse and interesting farming used to be is stamped all over the landscape. Just like the sheep, the cattle, goats, pigs and geese may well have been up on the hill during the summer months, both in order to take advantage of the huge amount of fodder on offer, but also to give the land on the valley bottoms a break while hay and other crops grew.

At High and Low Loup, the shepherds of Mardale had everything they needed to spend the summer months tending to their flocks in relative comfort. Drinking water would have been taken from any number of becks, or from nearby Rowantreethwaite Well, a mossy hollow from which the beck issues. The area of pasture surrounding the huts is lush and the animals would have been well fed, protected from the worst of the weather by the encircling hills and well watered by Captain Whelter and Hopgill Becks. Cranberries and bilberries could have been picked to vary the diet. Peat cutters and the foot traffic passing between Swindale and Mardale would have supplied news, company and provisions. It would have been hard graft, but I can think of worse ways to spend a summer.

With dinosaur hunting done for the day, Elliot and I packed away the camping stove and took the Old Corpse Road back down the hill, walking in the footsteps of the dead, the howl of ghostly wolves in our ears.

CHAPTER 6

Shadow Species

CATSTYCAM:
The steep path frequented by wild cats
(Middle English/Old Norse)

After the hay cut a few years ago, Becki and I took the kids camping in Swindale's fuzzy meadows. Following marshmallows toasted over a stick fire on a gravelly riverbank, our feet in the babbling water, we turned in for the night, collapsing into a sixteen-limbed tangle of discarded sleeping bags. After breakfast, we struck out on the snaking route towards the valley head, leaping flushes full of sundews and butterwort, mirrored by stonechats hopping along the fence posts.

Swindale Falls, or forces to use the Cumbrian vernacular, are, in my completely biased opinion, the finest in the Lake District. They're not the highest or the broadest, but their complexity and the plant life they support is breathtaking. On that still, sunlit morning with my family, it was their appeal as a place to bathe that had lured us. The water has carved at least half a dozen swimmable pools over the millennia. The best is the highest and the hardest to get to.

Passing by the lower pool, where mountain everlasting grows,

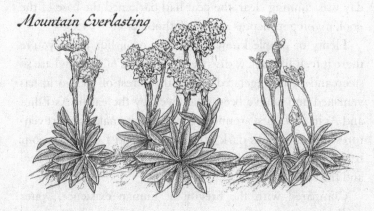

Mountain Everlasting

we followed the indistinct winding path through the bracken, taking us up the southern shoulder of the forces' deep gorge. Below us, ash, rowan, birch and juniper cast their shade over the white water as it churned past a bedlam of boulders. Way above the beck, as the terrain levelled off, we turned and dropped down into the steep-sided bowl that held the upper-most pool.

Having worked up a sweat in the August sunshine, we had no hesitation in discarding our clothes and throwing ourselves in. The water enters the pool at the back of the bowl as a 10-metre-wide wall of water, flanked by sheer sides clothed in ferns and flowers. A small rocky bay at the bottom of the slope that we entered by made a place to change and to store clothes and bags. Fine gravel backed up against boulders allowed an easy walk in through the shallows.

Mosedale, the wide valley above Swindale, which supplies the falls with their water, holds the largest expanse of blanket bog in the Lake District, and so the torrent often has a tea-colour due to the particles of peat that it has picked up as it drains through the saturated land. Even though the water that day was running clear, the peat had darkened the base of the pool, making its depths seem unfathomable.

Plenty of people know about this pool, but when you're there it feels like a new discovery. The sides of the bowl are so steep and richly vegetated it is as if the rest of the world has vanished and you've been swallowed by the earth. As Elliot and Aphra splashed around in the shallows, making short ventures into the pool's dark centre, Becki and I drifted on our backs, looking at the halo of sky above us, relishing the cool and the calm.

Compared with the brevity of human existence, water might have been cascading into this pool for ever. Cocooned inside the bowl's flowery sides, it felt like we were somewhere untouched by the ceaseless transitions that characterized the world above. These timeless pockets of land are enthralling, a sharp contrast to much of the rest of the ever-changing landscape.

The only constant is change, so said Heraclitus. This is a truth that applies as much to farming in the Lake District as it does to any other part of life. Until the Second World War, Lake District hill farms were very different to those today. Terry McCormick's 2018 book *Lake District Fell Farming* asserts that 'for over a millennium Lake District fell farming has evolved and strengthened. Its family and high-skill artisan culture has managed its economy, for most of those years, within ecological limits. Hundreds of decisions made over generations were guided by the imperative of sustaining the health of land as the most precious asset for family and community.' In recent decades, the subsidy-driven intensification of farming has breached these ecological limits, threatening the environment every bit as much as the viability of farming.

A hundred years ago, Lakeland farms were generally small and there were more of them. Most would be classed as small-holdings by modern standards. Farming wasn't even typically a full-time occupation until relatively recently. The production of livestock was simply one of the many things that people did alongside working in the mines, building walls, barns and houses, cutting peat and firewood, making charcoal, baskets and barrels, and doing any other job that survival demanded. I'm not looking back to these days with a rose-tinted view. Life for the residents of the rural Lake District a hundred years ago and before would have been seriously tough, a constant battle with the elements and a skilful balancing act of survival that we can't really comprehend today.

The evidence for how different things used to be is in the landscape. In Swindale there used to be eleven dwellings, most of which are now ruins. At least seven are recognizably farms, with barns attached. The most recently abandoned of these we call the Sycamore Barn, for the vast craggy tree that grows at its corner, raining leaves onto its slate roof. I spent a night of fire and revelry with some friends in the Sycamore Barn a while back. We call it a barn, but until recently it was most definitely a house. I don't know when the last residents left, but it's easy to picture how they lived. Just inside the five-foot-high doorway is a stone staircase, leading to a floor that has disappeared. The window openings give a view out across the valley head. There is an open fireplace and inglenooks, which made handy places to put our candles. A while back, a badger took a shine to the barn, gnawing through the corner of the wooden door and burrowing into the earth floor. A suspended platform, built for rat- and badger-proof storage, provided a place for the five of us to lay out our sleeping bags, dreaming to the swish of bats flying over our heads.

Swindale's schoolhouse and chapel, dismantled ahead of a narrowly averted plan to turn this valley into another reservoir in the 1930s, gives testimony to the fact that this was a self-contained community. Old names for fields – Miley's Meadow, the Cow Pasture, Potato Field and Hoghouse Field – provide a fragile link to past land uses. These names don't appear on any maps, having been handed down by the valley's residents through the ages.

One of the things that worries people most about change in the Lake District is what it might mean for how the place looks. The Lake District is undeniably beautiful. Fortunately, for a few hundred years, artists, poets and writers have thought the same way and made a record that we can compare with today. Being something of a backwater, Haweswater didn't attract the attention of artists as much as the more famous parts of the Lake District did, so to gain a window into the past we must look elsewhere.

In 1824 the artist John Glover set up his easel near his home at Blowick Farm on the eastern shore of Ullswater. The result, *Ullswater, early morning*, is a painting that you could walk into. Unlike many more interpretive landscape paintings of the same period, it appears incredibly accurate. Several species of plants and trees are identifiable, and his rendering of the shapes and the positions of the mountains is perfect. The jagged teeth of Helvellyn and Catstycam mark the furthest horizon, with Arnison Crag and Glenridding Dodd closer to.

The painting depicts the scene on a still day in autumn. The birch trees are just starting to turn, but the bracken is already brown. The foxgloves are swollen after flowering, their leaves converting the dwindling hours of sunshine into sugars to see them through a winter of dormancy. Ullswater's surface is mirror calm, reflecting the wooded hills along the shore. Circular ripples show where fish have been rising. A small sailing boat is making its way north towards Pooley Bridge. In the foreground, four cows rest in a shady patch of bracken, with three sheep just behind them, grazing at the lip of a grassy bowl. The lakeshore villages are small and inconspicuous, but Patterdale Church is visible in the left of the painting through the branches of a straggly willow.

I went out in search of the spot where Glover sat to see how much had changed over the course of two hundred years. The ease with which I found it is a mark of the precision of his painting. Here was the boulder where the willow used to be, the roof of Patterdale Church visible beyond. Walls had been built in the intervening two centuries, and the trees had grown, partly obscuring the view of the lake, but it was unquestionably the same spot. I'm not much of an artist, but with a printout of the painting to hand, I took a photo framed to match the view as closely as possible.

The most striking thing about the comparison between the landscape today, snapped with my phone, and the one that Glover rendered in oils two hundred years earlier, is just how similar they are. The mountaintops are still largely bare, the lake margins still mostly wooded. They are unmistakably the same landscape, full of distinctive rugged Lake District charm.

And yet, if you could inhabit the painting, listen to its soundscape, look more closely at its unpainted detail, sharp contrasts would be revealed.

In 1892, sixty-eight years after Glover completed his painting, the Reverend H. A. Macpherson published *A Vertebrate Fauna of Lakeland*. The Reverend was an expert naturalist, as many men of the cloth of his generation were. His book provides detailed accounts of all the mammals, birds, reptiles, amphibians and fish, both extinct and extant, that were found in the counties of Cumberland, Westmorland and Lancashire north of the Sands, which is conveniently close to the extent of modern Cumbria. For someone as obsessed with the nature of the Lake District as I am, his book is essential reading.

Most of the species that Macpherson writes about are still here and, although he doesn't provide numbers, his book suggests that many that are currently struggling, such as curlew, lapwing, golden plover and grey partridge, were doing much better in his day. The most compelling accounts, though, are of those species that aren't with us any more, and their list is depressingly long. The more I read, the more robbed I felt.

Although our meadows in Swindale have endured, they are missing something. A hundred years ago, they would have been home to corncrakes, blackbird-sized dry-land relatives of coots and moorhens. Because meadows were so much more widespread than they are today, corncrakes were common as muck, so common in fact that according to the Reverend they even occasionally nested in people's gardens. Up until the 1960s their distinctive nocturnal call would have been a familiar feature of farmland up and down the country. Accounts from the time give no hint that their numbers were soon to take a dramatic nosedive.

Corncrakes are secretive, more often heard than seen. Their repetitive raspy call, which they make most during the night, sounds like a pencil being dragged across the teeth of a comb and earned them their onomatopoeic Latin name, *Crex crex*. They are most vocal soon after they arrive from their African wintering grounds in April, making their call a sleep-shattering 20,000 times per night.

Corncrakes rely on a degree of general untidiness. When they first arrive on their breeding grounds, they need to find tall vegetation in which to hide straight away, but hay meadows haven't usually grown to a height to give them the cover they need by then. While they await the growth of the crop, corncrakes make use of beds of nettles or irises, which grow in the scruffy, wet and weedy parts of the landscape that have in so many places been swept away, drained and turned into something agriculturally more productive. As traditional hay meadows leaked out of our landscapes, these sanctuaries of colour, nectar and insect life that used to be undisturbed oases for corncrakes to rear their chicks became sterile green deserts, through which tractors ripped two or three times a year. The corncrakes didn't stand a chance. Sixty years on from Macpherson's death, other than sporadic breeding attempts by the odd pair, they were extinct in England.

In the extreme north-west of Scotland and Ireland, where traditional farming practices are being kept alive, corncrakes cling on. There is also a small reintroduced population in the Midlands, where the RSPB and others are working hard to return them to at least some of their former haunts.

There are lots of good hay meadows in the Haweswater area, extending from ours in Swindale down the valley as far as Bampton. Many of these are traditionally managed, full of hay rattle, oxeye daisies and bistort. I like to think that maybe one day they could accommodate corncrakes again. During our night in Swindale with the kids, other than the odd hoot from a tawny owl and the almost subsonic clicking of the bats, the meadows were eerily silent. No corncrakes disturbed our slumbers. I'd happily trade a good night's sleep to have them back.

Although I can't find any that relate to corncrakes, there are countless places in the Lakes that carry the name of a now locally extinct species. Looking at the framed Ordnance Survey map of the area around our house, which hangs in our spare room, one jumps out at me immediately – Cocklakes. This is a derivation of 'cock lek', meaning that at some point in the past this undulating farm, five minutes from home, was very likely to have been the venue for one of the UK's most impressive natural spectacles.

Black grouse, also known as blackcock, have disappeared from the Lakes but not from Cumbria altogether. I went to visit them last spring. Setting out from home with a flask of coffee in the early morning dark, I headed for RSPB Geltsdale in the North Pennines. Geltsdale is a huge reserve, twice the size of Haweswater. It's well off the beaten track, and the tiny car park at the end of a bumpy dirt road is a good indication of how few visitors make the effort. More should – it's a wonderful place. The team at Geltsdale have been managing their land for much longer than we've had our tenancies at Haweswater, so I visit quite regularly in order to learn from them.

Even though they are just across the Eden Valley from the Lake District, the Pennines are a completely different landscape. A long chain of huge hills running up the spine of northern England, much of the Pennines' rolling geology is dominated by blanket bog, almost unbroken by crags and rocky knolls. As a result, the Pennines, particularly their northern end, feel more open, windswept and bleak than the Lakes. They also have a lot more birds. Golden plover, curlew, lapwing, redshank, short-eared owl, hen harrier and red grouse, all few and far between in the Lakes, fare much better in the Pennines.

I don't think I can recall a single time when I've been to Geltsdale without it raining. Pulling on my boots in the early morning mizzle, I could hear curlews and skylarks trying their best to sing up the sun. I was here to help my Geltsdale colleagues with their annual black grouse survey. With snipe drumming over the peaty ground, I made my way up the hill until I reached my position.

Every spring, black grouse congregate in traditional sites to 'lek', a spectacular ritualized tournament where they compete for mating rights. Maintaining a respectful distance so as not to disturb the action, I had a clear view through my binoculars of an area of short grass, where four male grouse appeared as the damp sun came over the hill.

Watching a lek is an inimitable and somewhat voyeuristic experience. The males have bright red wattles over their eyes like cartoon eyebrows, which, set against the glossy black of their heads and bodies, give them the look of Marvel supervillains. The word 'lek' is another gift from the Vikings, meaning 'to play', though a lek is much more than a game. Puffed up to maximum size, with lacy white tail feathers on full display, the male grouse strut and lunge at each other, all the while sustaining a trilling, bubbling coo, interspersed with a strange mechanical hissing. They are truly one of our most beautiful birds, and watching them lek, I got the sense that they knew it. While the males perform, the dowdy females, known as greyhens, lurk in the longer grass at the edge of the ring. Like me, they stay hidden while they eye up the talent. The most impressive males win mating rights to the females. After an hour or so, the show was over, and the birds vanished into the landscape to mate in secrecy. I walked over the hill to the reserve office to add my count to the group's tally.

Black grouse are one of many shadow species in the Lake District – we see the signs of their presence everywhere, but the animal itself is missing. The last black grouse disappeared in the 1980s, making them one of the most recent departures from the national park's cast of breeding birds.

They demand a lot from their environment. At different stages of their life black grouse need intact bogs and flushes, heathland, young woodland, and grassland to satisfy a diet that includes insects, berries, and the buds and shoots of various trees and dwarf shrubs. Their disappearance from the Lake District, as well as many other parts of the UK, is an accidental consequence of the way that we've looked after our countryside over the course of the last few decades. We have gradually been chipping away at the habitats that the black grouse and other wildlife need. Heaths have been planted with conifers, bogs drained, scrub and long grass cleared away.

The tide is beginning to change, though. Thanks to the combined efforts of farmers, landowners, charities and government agencies, positive habitat restoration is happening all over the national park. Large areas of new native tree planting have been established in recent years, some of which could help to create a habitat corridor between the Lakes and the parts of the Pennines where black grouse are still present. The area around Haweswater is the last place they were found in the Lakes, and I hope it's where they will recolonize first.

If black grouse return one day to strut their stuff in the Lake

District, it will signal that we have turned a corner in our relationship with the area's wildlife. I'd love to see them back at Haweswater for their own sake, but also because they could be prey for the birds that I want to see return more than any other – the eagles.

Golden eagles will always be part of the DNA of Haweswater. Whenever I give talks about our work, the first question I'm asked afterwards is almost always about eagles and whether we are planning to reintroduce them. I'd love to be able to say yes, but the truth is that we aren't ready for them yet. Our focus is on improving habitats in the hope that one day they'll arrive under their own steam, as they did before. This would show that the landscape is fit for them again.

Before their 1969 return, Lakeland had been without breeding eagles for at least 170 years. But there is ample evidence that they belong here, and not just one species of eagle either. Glance over any detailed map of the Lake District, and it won't be long before you spot an Eagle Crag. There are also several Heron and Iron Crags. Herons don't nest on crags, and many of the iron crags are nowhere near mining areas. Both names are corruptions of *erne*, an alternative name for the white-tailed eagle. Haweswater has an important place in the history of these even larger eagles too. The first trustworthy account of a white-tailed eagle eyrie in Cumbria was from 1692, on Wallow Crag, which stands woody and imposing overlooking the dam, less than a mile from my desk at Naddle.

When Reverend Macpherson published his book in 1892, both eagle species had already gone, but his long and grisly accounts are packed with details from people who had lived alongside them. The locations of eagle nests were generally well known, predominantly in order to aid the theft of their eggs and the killing or capture of the adults. Macpherson describes harrowing episodes of the destruction of both adult birds and eggs, including an account of the eggs from that same Haueswater nest on Wallow Crag being robbed in 1787. There were occasional sightings of white-tailed eagles for a few years after that, but no more of these sky giants bred in Cumbria beyond 1800. By 1920, there were none left in the UK.

White-tailed eagles, also sometimes referred to as sea eagles, are as much at home on the coast catching fish as they are further inland. This adaptability partly explains why white-tailed eagles historically outnumbered golden eagles considerably. Golden eagles are birds of wild and lonely places, shunning people wherever possible. This made them harder to find and hunt down, so they persisted in remote parts of Scotland, while the bolder white-tailed eagles were exterminated everywhere.

Both eagle species have had a torrid time over the last 200 years, but in recent decades their fortunes have started to improve. Thanks to a long-running reintroduction programme in Scotland that began in 1975 using birds from Norway, white-tailed eagles have returned. Further releases of these flying barn doors on the Isle of Wight should help them achieve nationwide re-establishment in the not too distant future. Spike and a group of volunteers were lucky enough to see one of the Isle of Wight birds circling overhead while working in Rigindale recently. We'd like to think she was checking it out as a future breeding site. Because they roam so widely in their early years, there's now a chance of seeing a white-tailed eagle almost anywhere. Their 2.5m wingspans make them hard to miss.

With full legal protection, golden eagles are also on the up with numbers exceeding 500 pairs in 2015. They are almost all north of Scotland's central belt, bar a few pairs in the Borders. A project to boost this small and vulnerable southern population has been translocating young golden eagles for the last few years, taking chicks from nests in the Highlands and rearing them by hand. Golden eagles often lay two eggs, with one usually hatching before the other. The younger of the two is at a disadvantage and is often killed by its older sibling – nature is rarely kind. Taking one of the two eggs for captive breeding and release to bolster a population elsewhere therefore has little impact on the area from which it was taken and can increase overall chick survival.

If the southern Scottish project goes according to plan, then there should be a much healthier population of golden eagles looking to set up new homes on the Lake District's doorstep. As far as an eagle is concerned, the Lake District is part of the same landscape as the Scottish Borders; they don't see a national divide. If we can make the habitat right for them, it's more than possible that they will grace the skies above Haweswater again.

Their return might not be universally welcomed, though. One of our neighbouring farmers told me a while back that if eagles reappeared at Haweswater he would take matters into his own hands to ensure that they didn't stick around. I never really got the sense that there was much resentment towards eagles while we still had them, so this threat took me by surprise.

Nervousness from sheep farmers about eagles returning to the Lake District is perhaps understandable, since eagles certainly consume dead sheep and, if the opportunity arises, they can also take very young lambs. That said, current sheep-farming practices wouldn't give them the chance very often.

Today, sheep are closely tended to at lambing time and lambing often takes place in barns. Ewes with young lambs are usually kept under close watch while they are at their most vulnerable, in the fields closest to the farmhouse. These are not places that golden eagles would come anywhere near, and with dogs, farmers and vehicles around every corner, they are likely too busy for white-tailed eagles too.

On the Isle of Mull, where reintroduced white-tailed eagles are now well established, they do take live lambs – a fact which was widely disputed for many years – but in very small numbers, contributing a tiny proportion of the total typical lamb mortality. The fact that only small numbers are taken isn't much comfort to a farmer, though. Every lamb is a life that they have helped to bring into the world – just because the numbers are low, doesn't make lambs lost to eagles any easier to stomach.

Despite their depredations, the eagles of Mull also have an economic upside. They've spawned an eagle tourism industry which brings in £5 million of annual spending and supports 110 jobs; significant numbers in a small island community. Balancing the impact on sheep farming against wider economic and ecological benefits is a tricky business and is brought into sharper focus when the species has returned as a result of conscious reintroduction efforts, rather than natural spread and colonization.

By whatever means, it's surely only a matter of time before we'll see eagles back in the skies of the Lake District again, hopefully of both species. It won't be plain sailing, and they won't be welcomed by everyone. I just hope that they choose Haweswater, as they did once before.

The eagles are conspicuous by their absence. Other birds made their departures more quietly. Corn buntings, small brown birds with a song that sounds like jangling keys, disappeared from the Lakes in my lifetime, along with the small arable plots they depended on that used to be a feature of Lake District farms. Yellow wagtail populations crashed around the same time in response to the drainage of damp pastures and a move away from cattle grazing. Red-backed shrikes, common in Cumbria a century ago, no longer breed anywhere in northern England thanks to the demise of insect-rich scrub and hedges. Wrynecks, woodlarks and nightjars all vanished from the Lakes in the mid-1800s, as a result of the general tidying up of the countryside, and the loss of heaths, woods and wood pastures.

Mammals, too, have been disappearing. Pine martens were hunted to extinction in the early 1900s; wildcats went the same way a few decades earlier. Both species are commemorated in the names of crags all over the Lake District. Water voles were wiped out more recently due to the loss of their watery habitat and predation by invasive non-native mink.

It's not quite all doom and gloom, though. Macpherson's entry on red kites charts a decline to extinction in Cumbria by about 1840. Now, red kites are back, thanks to a UK-wide reintroduction programme, and their fork-tailed silhouette is being seen in Lake District skies with increasing regularity. We had a kite roosting in Naddle Forest a few years back, and we're keeping our fingers crossed that they'll start to breed again one of these days.

Ospreys have returned, and, like the kites, their numbers are also increasing year on year. Knowing that they were absent until so recently makes catching sight of one pulling a fish from a lake all the more thrilling. Buzzards, peregrines, goshawks and hen harriers, while still suffering from illegal killing in some places, are all up from historical lows caused by sustained persecution.

Despite these few bright spots, the general picture is one of decline and loss. The fate of most of our shadow species is wrapped up with the changing ways in which we have looked after the land that we once shared with them.

❧

Using place names, paintings, books and stories, it's easy enough to reconstruct the general picture of the Lakeland landscape, but understanding the specifics is harder. The further back you go, the hazier the details become. One important feature that's virtually impossible to determine from paintings or other historical sources is the abundance of flowers. At the perspective that John Glover was painting, he depicts some of the bigger plants – the foxgloves, bracken and trees – but the composition of the sward at his feet is rendered down into a colourful textured pattern. It's tempting to try to interpret his representation as hinting at greater diversity than I'm used to seeing now, but that could easily be wishful thinking.

Historical accounts of the Lake District's flora tend to record a species' presence rather than its abundance. All they provide is a list, and there's often no way to infer just how widespread each species might have been. There are some useful sources that shine a light, though. Cumbria Wildlife Trust carried out an oral history project a few years ago, interviewing people who could recall the ways the land was looked after before intensive farming took hold. Stitched together into a series of short films, the recollections of a way of life so recently passed, spoken in the tremulous voices of those who lived it, are haunting and heart-breaking.

What comes across more than anything is what a flower-dominated world they inhabited. Flowers are mentioned again and again, either as ubiquitous features of the hay meadows or as items for harvesting. One man recalls collecting wood crane's-bill for his class to study for O Level exams; in another film a woman talks about collecting the edible root nodules of pignut, crimson rosehips and nettles to sell. The rich scent of hay and the colours of flowers across a landscape are palpable. That people still alive had once lived in a world so much more vibrant than today moves me; it's like some glorious treasure that's just beyond reach.

Everything I've read, heard and seen convinces me that the Lake District landscape before the 1950s had a lot more flowers in it, and it's beyond doubt that the intensification of farming that occurred in the post-war decades has been a major contributor in their decline. However, since the 1990s things have started to improve. I'm seeing hay meadows being restored all over the place. The work we're doing in Swindale to breathe life back into our meadows is one of many similar initiatives. Sheep numbers are falling to more sustainable levels, and I know many farmers who are returning to more mixed systems, with cattle, pigs, poultry and ponies. There are more trees in the landscape now than there were in the 1950s. Granted, much of the increase has been down to conifer plantations, but more native broadleaf trees are now appearing too.

There is hope in the fact that so much change has happened in such a short space of time – if the pendulum can swing so fast in one direction, maybe it can swing back as swiftly. The vanishing generation, giving voice to their memories over sepia photos, is a fragile link to a flower-rich past. It's good to know that some of that richness is being restored in time for them to appreciate it.

❧

The Lake District, with its constantly evolving farms and an ecosystem full of holes, became a World Heritage Site in 2017. Two earlier bids had been made back in the 1980s, but they didn't align with UNESCO's categories at the time so weren't successful. With the creation of a new 'Cultural Landscape' category in 1993, a third attempt was made. The development of the successful bid started in 2001, in response to the outbreak of foot-and-mouth disease, which took a major toll on farming and the wider economy in the Lakes. The intention was to cement the Lake District's reputation as an internationally renowned destination to help support tourism and rural communities. The fact that it took fifteen years to achieve the consensus needed to finalize the bid probably should have set alarm bells ringing.

While the bid was being prepared, those of us seeking the recovery of nature in the national park began to worry that a World Heritage Site badge, far from helping support the landscape, might be a barrier to change. Early versions of the application documents included commitments to stop tree planting and to prevent further reductions in sheep numbers, and so our concerns grew.

The RSPB are members of the Lake District National Park Partnership, a group of twenty-five organizations that try to steer the management of the Park. It is this Partnership that applied to UNESCO, and so the RSPB is partially responsible for the designation. There are many vocal critics of the Lake

District becoming a World Heritage Site, and we've taken some flak for not opposing it. As ever, things are more complicated than they might seem from a distance. I have no doubt that if we'd tried to block the application, or walked away from the partnership, the designation would have happened anyway. Together with the other environmentally minded members of the partnership, we challenged some of the most potentially damaging elements of the bid and have ended up with a World Heritage Site which is far less restrictive than it might have been. This is a recurring theme of our work in the Lake District, and in conservation more generally: no matter how uncomfortable it might feel, compromise and working with those with different priorities is necessary in order to ensure that nature's voice is heard.

The nomination document that describes 'The English Lake District World Heritage Site' runs to 716 pages, so it's difficult to summarize what the designation really stands for. It talks a lot about beauty, about farming and sheep, especially Herdwick sheep, the Lake District's native breed beloved of Beatrix Potter. It mentions nature now and again and talks about the Lake District as the birthplace of the conservation movement, though it means landscape conservation, which focuses on preserving the aesthetics of a place, rather than nature conservation, which is more concerned with the protection of species and habitats. It celebrates the area's geology and pretty lakeshore villas, its poets and farmers. If you look hard enough, you can find sections that support or oppose almost every possible point of view, but the emphasis on sheep farming is clear. The word 'sheep' appears 365 times, the word 'flower' only three times. 'Farm' appears 1,052 times, 'nature' ninety-two times.

I'm sure that there are lots of people who care passionately about the designation, but I've not met many of them. Most of the farmers I know are ambivalent – contrary to what you might expect, World Heritage status doesn't provide them with either funds or protection. The Federation of Cumbria Commoners initially welcomed the designation, having described it as a 'powerful weapon that puts hill farming centre stage', but it's not clear how that weapon is to be used.

A couple of years after the inscription, a delegation from the Lake District World Heritage Site Steering Group came to Naddle Farm to give my RSPB colleagues and me a training session, designed to help us understand what being in a World Heritage Site meant for our conservation work at Haweswater and elsewhere in the Lakes. They talked us through the cultural concepts from the nomination document, and the new paperwork we now had to complete to enable the steering group to ensure our activities didn't impact on the World Heritage Site's attributes.

In the discussion afterwards, I asked how the RSPB's presence at Haweswater was perceived from a World Heritage perspective. I was told that when the application was being prepared for UNESCO, the steering group had been forced to accept that there were a number of 'warts' on the face of the potential World Heritage Site. One of these warts was the RSPB's presence at Haweswater. For my and my colleagues' work to have been described like this was extraordinarily offensive, and the fact that anyone would be prepared to say something so blunt to our faces took my breath away.

I asked for clarification, just to make sure I'd heard correctly. It was explained to us that we weren't 'authentic'. Because we didn't fit the stereotypical profile of family farmers, we were considered second-class citizens. They were effectively saying that the only true stewards of the land in this newly minted

World Heritage Site were the Lake District farmers, ideally born and bred here and from farming families, lineages to which my colleagues and I didn't belong. Yet they also worried about the lack of new entrants to farming. It's hard to see how this sort of thinking can end well.

Proponents tell me that the English Lake District World Heritage Site is a celebration of the beauty and harmony resulting from the interaction of the Lake District's people and its landscape. But beauty and harmony to who? A beautiful landscape to me is one with no straight lines, plenty of colour and humming with life. To others, neatness and order nourish the soul. Where some might hear harmony, others hear discord. The Lake District is always changing, as are the myriad ways that we interact with it. Perceptions of its ever-moving tapestry vary from person to person, from moment to moment. Trying to capture the indefinable essence of a place is like trying to catch the wind.

I love the Lake District, but it's a love tinged with an overwhelming sense of loss. I pity the hillsides without their flowers and trees, and I miss the farming that worked with nature. I crave the return of stolen riches. That UNESCO have decided that a place so ecologically wounded is of 'outstanding universal value to humanity' perplexes me. I understand that the World Heritage Site designation isn't about nature, but culture and nature are inextricably linked. Without a healthy, functioning ecosystem, farming and its cultural associations will wither and die.

Now that Haweswater is part of a World Heritage Site, there's more paperwork for us to do. For any project that involves change to the landscape, be it building a hide or restoring a river, I'm obliged to complete a Heritage Impact Assessment. Being in a World Heritage Site means that the threshold for planting trees without an Environmental Impact Assessment screening has dropped to zero – if I want to plant a single tree, I must consult the Forestry Commission. Because the World Heritage designation covers the entirety of the national park, it's now easier to carry out conservation work outside of the Lake District than inside it. I'm sure that's not what the original architects of our national parks would have wanted.

You'd be forgiven for thinking that wildlife would be well protected in a national park. Sadly, this isn't the case in the UK. Assessing the condition of nature across a given area isn't easy, and the paucity of good data is a problem in itself. The most widely used metric is the condition of Sites of Special Scientific Interest (SSSIs), the nation's network of protected areas, which represent the best of our habitats and species. According to the Lake District National Park's 2018 State of the Park report, only 21.6 per cent of SSSIs in the Lake District are in a favourable condition.* Many worry that World Heritage Status may put an additional strain on wildlife already in dire straits.

A World Heritage Site designation comes with no money – it is simply a badge. The tourism sector may well benefit through the increased numbers of visitors that the badge has pulled in, but I doubt much of the extra income generated makes its way into farmers' pockets, so it's hard to see exactly how World Heritage is helping to sustain the farming system on which it is predicated. Thankfully, if a World Heritage Site is deemed to be damaged in some way during an inspection,

---

* The picture is similarly bleak for national parks in England as a whole. Only 26 per cent of SSSIs are in a favourable condition inside England's national parks, compared with 43.5 per cent that meet the same standard outside (Cox et al., 2018).

the only punitive measure that UNESCO have is to take away the designation.

If the Lake District World Heritage Site was about celebrating, protecting and sustaining a farming system that genuinely worked in harmony with nature, then I'd be all for it. I'd love to feel proud of living and working in a World Heritage Site where farming and wildlife benefited each other, a place that was as nature-rich as it was culture-rich. Whether World Heritage will help or hinder this becoming a reality, or have no impact either way, only time will tell.

When the Lake District became a national park in 1951, its streams and ditches were busy with water voles, corncrakes rasped from the meadows, black grouse lekked on scruffy fellsides and corn buntings perched on fences around little arable plots. John Glover's 1824 rendering of Ullswater, so visually similar to the scene today, had even more wildlife hiding in it. While he sat painting, he might have been distracted by yellow wagtails picking insects from the cattle dung, red-backed shrikes impaling grasshoppers on the thorny scrub or wrynecks twisting themselves in snake-like contortions. As Glover packed up his easel, nightjars could have been churring in the failing light. Walking home through the woods, skulking wildcats and pine martens might have eyed him warily from the trees as he passed.

Wordsworth, born in 1770, celebrated a Lakeland that was every bit as beautiful as it is today. His Lakeland, though, had golden and white-tailed eagles soaring over it.

When the great walls that encircle Naddle Forest were first set out in the 1300s to create a preserve for the wealthy to hunt in, wild boar would have been on their list of quarry. Wolves still stalked the fells, cranes trumpeted from the bogs and beavers engineered tangled wetlands full of fish in the valley bottoms.

The Lake District is impoverished in many big and obvious ways once you start to look; important pieces of the ecological jigsaw are missing. We've become accustomed to the Lake District in its current state, used to seeing the hills covered in a mat of close-cropped green and little else. Few people appreciate that this landscape is an artefact of our management, and that it hasn't always been this way. Visit sensitively managed comparable mountain landscapes elsewhere in Europe and you'll find yourself walking through carpets of flowers, humming with life.

Diverse and abundant plant life supports a complex web of other organisms. There's no point throwing a few pairs of golden eagles back into the landscape if it won't provide them with anything decent to eat. Though, when the time is right, giving some of our missing species a helping hand to come back may well be necessary. The science of reintroductions is a rapidly developing field, and there are an ever-growing number of successful projects that have returned previously extinct species to their former haunts in the UK and globally. Beavers are back, both in free-living populations and in many large enclosures across the country. Red kites, goshawks and white-tailed eagles are expanding, all thanks to direct human intervention.

Some species are returning under their own steam. Pine marten populations are expanding in Scotland thanks to reduced hunting pressure, and sightings in Cumbria are increasing as they slink south across the border. This could spell good news for red squirrels. Having lived alongside pine martens for millennia, red squirrels have evolved tactics that help them evade predation, which the non-native grey squirrels have not. In places where pine martens have returned, grey squirrel numbers have tumbled.

Others may need more help. Black grouse are famously poor at dispersing into new areas of suitable habitat, so even with effective habitat corridors they might require reintroduction to help them retake lost ground. Corncrakes probably wouldn't find their own way either, and wildcats would need a captive breeding and release programme to enable them to stalk our craggy woods again.

For any of these species to thrive, the landscape must be fit and ready. We need to take a whole-ecosystem approach, addressing ecological function and the connections between our myriad species and the habitats in which they live. There's a lot of work in all this, plenty of potential employment and activity before we get to a state where nature will be better able to sustain herself.

At Haweswater, we are starting with the flowers and working our way up. Plants, and flowering plants in particular, are too often overlooked. We need to learn to appreciate wildflowers more, to notice and nurture those we have, and to restore the colour to our landscapes.

If we do that, we'll have more insects, more birds, more reptiles and mammals. Achieve that, and we'll also have a landscape with more wildness and wonder, more worthy of being passed on to our children.

A landscape of flowers is a landscape of hope.

PART 2

Inspiration

Come forth into the light of things,
Let Nature be your teacher.

William Wordsworth, from 'The Tables Turned', 1798

CHAPTER 7

Across the Border

CARRIFRAN GANS:
The fort of ravens
(Celtic)

It is an unassailable statement of fact that flowers and plants are the basis for all life on earth. If it weren't for plants and their ability to convert the energy of the sun into sugars, the earth would still be a cold dead place.

There are 391,000 species of vascular plant on the planet, distinguished from the more diminutive mosses and liverworts by having tissues that transport fluids and minerals through their bodies. Vascular plants include such diverse groups as ferns, clubmosses, horsetails, cycads and conifers, but the largest group by far are the flowering plants, or angiosperms. Perhaps confusingly, grasses and their allies, sedges and rushes, are also classified as flowering plants, but they only number around 20,000 species. Plants with true flowers, which include broadleaf trees, make up the majority of all the vascular plants on Earth. Without them, the planet's sprawling web of life, on which all creatures depend, would disintegrate.

The visual variety of flowers hints at a far greater chemical diversity within. Every species has a unique combination of organic compounds, bubbling away in their tissues. We benefit from this biochemical apothecary in incalculable ways. Flowering plants supply us with an endless variety of food, medicine, cosmetics, dyes and building materials. New species of flowering plant are discovered every year – who knows what as yet unknown benefits they might provide us with.

Leaves and stems feed all sorts of herbivorous creatures, but flowers offer up a particularly special treat. Nectaries secreted within flowers provide butterflies, moths, wasps, bees, flies, beetles, ants, hummingbirds, bats and the occasional lizard and monkey with a sugary reward. Flowers support a staggering array of other life.

They look pretty in a vase on the kitchen table, but stitched together to clothe a vista in their glory, flowers can transport you to a higher plane. Add in all the other wildlife that a landscape of flowers supports and you have a place of great visual and sensual beauty, a landscape to soothe and heal bodies and minds ground down by the hardships of modern life. This is an experience hard to have in the UK.

Although there are areas of intact habitat at Haweswater, the bulk of the land we inherited was a blank slate, reduced to species-poor acid grassland. A big part of my role is about effecting habitat change, working with others to turn these spaces into something richer.

A plan for restoration needs a clear sense of what the end result will be. A sensible starting place is to understand what habitats were here in the past. Drained areas of peat soil would have been bog dominated by sphagnum moss. The ghosts of channels tell where rivers used to run. Remnants of woodland ground flora or a covering of bracken indicate that an area once had more trees. Reading the landscape like this allows you to reconstruct it in your mind's eye, but that can only teach you so much. To really experience intact upland landscapes, to get to know how they smelt and felt, I needed to travel.

⬟

Philip Ashmole is a truly inspiring octogenarian. A zoologist educated at Oxford, he spent most of his scientific career studying the bird life of tropical and sub-tropical islands. He has discovered several species of spider new to science and developed a widely accepted concept that describes the foraging behaviour of seabirds, known as 'Ashmole's Halo'. For me, as important as this research is, his place among the angels will have been earned for the work he has done later in life in the Scottish Borders.

The drive north from Cumbria through the Borders on the A74 towards Glasgow is about the most depressing trip an ecologist can make. Miles and miles of conifer plantations have obliterated swathes of the moors, mires, woods and meadows that were there before. Between the plantations, the impacts of drainage and intensive grazing are plain to see. The Lake District has its fair share of ecological problems, but the pastoral softness of the valley bottoms juxtaposed with the soaring rocky crags at least gives it a visual charm. Without these, the view that most people get of the Borders is a landscape of real severity.

This is the region where Philip and his wife Myrtle, a couple with a deep love of wildness, chose to live out their retirement. After working and studying nature in many countries where it's still thriving, it must have been jarring to return to the UK, a nation placed in the bottom 12 per cent of countries for 'biodiversity intactness', and even harder to come somewhere like the Borders, where what wildlife there is has been forced into such tiny fragments. Philip and Myrtle aren't people to sit back and accept the status quo. Their house today sits surrounded by woodland and wetland that they created, a wildlife oasis where once there was just bright green farmland. Spurred on by these local successes, the desire to do more than just enhance their immediate surroundings began to grow. In 1996 they and a group of like-minded friends established the Borders Forest Trust and started on a plan to breathe life back into much more of their landscape.

The trigger for the Trust's formation was the knowledge that Carrifran, a 660-hectare valley a few miles north-east of the market town of Moffat, was coming up for sale. On the face of it, Carrifran was nothing special. A place of rock and clipped grass, a glacial valley stripped to its bones, like countless others in the area. Like Haweswater, Carrifran's valuable habitats had been reduced to stubs, clinging to life on the margins, beyond the reach of animal teeth. The valley's one rowan tree, on the edge of the Carrifran Burn*, was the sole survivor of a lusher past.

Carrifran falls within the Moffat Hills Site of Special Scientific Interest, a designation which gives it a degree of legal protection. The document that explains why the area warrants safeguarding states: 'Acidic grassland comprising bent *Agrostis* spp, and fescue *Festuca* spp, grasses ... cover much of the site but in damper peatier areas heath rush *Juncus squarrosus* and mat-grass are widespread.' This is ecology-speak for 'catastrophically overgrazed'. The species listed here are the coarse unpalatable ones that dominate when grazing has eliminated all the tastier ones. The site's main importance is

---

* A burn is a large stream or small river in Scotland.

its geology, which thanks to centuries of heavy grazing had been comprehensively exposed. The citation also mentions Carrifran's remnant of Arctic-alpine flora, hanging on around a series of steep waterfalls near the valley head. In case the sheep and the deer hadn't made life hard enough for these botanical survivors, the area also had a population of feral goats, able to access terrain that was even more vertiginous.

Philip, Myrtle and their friends pieced together the past, studying archaeological evidence and pollen cores that gave a window into how vegetation had changed over time, helping them to see what Carrifran had been and what it could be again. The plan they developed was for a living valley restored, re-wooded, re-flowered, full of birds and other wild creatures.

Buying a valley is no small feat for a fledgling community-based charity with no financial reserves, and the first few years of the Trust's existence was all about raising funds and supporters. Despite the challenges, the clarity of their vision and the strength of the characters that believed in it carried them through, and on 1 January 2000, Carrifran became the property of the Borders Forest Trust and its transformation began.

First, the root cause of the valley's bald state had to be addressed. The sheep flock that had grazed here previously had gone with the outgoing owner, but neighbouring valleys were still used as pasture, so fences were erected to stop other sheep wandering in. The feral goats were shot, as were deer, ensuring their number stayed low enough to allow trees and plants to recover.

Then came planting, lots and lots of it: more than half a million trees. Because Carrifran had been treeless for so long, the potential for natural regeneration was almost non-existent. There were no native woodlands in the immediate vicinity, so there was very little chance that seed would be blown in or be brought in by birds. Without a viable seed source, and because the many supporters of the project wanted to see meaningful change in their lifetimes, the decision was made to plant the trees, rather than wait the decades or maybe even centuries it might take for them to arrive by themselves.

Tree planting is what occupied armies of local volunteers for the following fifteen years. Tree seeds were sourced as locally as possible, lovingly grown in nurseries and then brought to their new home. The planting was carried out with astonishing attention to detail. Nowhere are there straight lines. Willows and alders are by the burns and in other damp places; hazel and oak are where bracken was growing, marking out the deeper soils; montane willows stud the higher reaches; the mires and screes remain open. To visit Carrifran today, it's difficult to believe that the tree cover that exists is anything other than completely natural. There is a lesson here. Too many new woodlands have been planted without the same depth of ecological understanding. Too often they end up as ugly ranks of seemingly random species, unmatched to ground conditions, still wearing plastic tree guards years after they should have been removed.

On my first visit to Carrifran, a party of us from the Lakes were shown around by Philip and other Borders Forest Trust founder members. Hearing the story of the valley's restoration from the people that did all the hard work was an inspiration, their deep knowledge of the place and their connection to it was evident in every sentence. Philip was representing a community of people who had ploughed time, effort and love into this valley; they have become true custodians, and their pride will grow with the trees that they have planted.

After making our introductions in the small car park at the valley bottom, surrounded by willows, rowan and hazel, we set off up the valley. For a man well into his eighties, Philip is impressively spry; many of the members of our party half his age struggled to keep up. Watching and listening to him here in his element, we got the sense that his whole being was sustained by his enthusiasm and passion for his work. The sparkle in his eye and his constant smile were those of a man who knows he's made a real difference.

While the most striking change at Carrifran is the massive increase in trees, there are many more subtle transformations too. Both the number and variety of birds in the valley has soared since the new woodlands were planted, with at least nineteen species having colonized. Periodically, the young woodland gave way to glades where soils were too wet for trees. These are now full of tall wildflowers; meadowsweet, valerian, lousewort, greater bird's-foot trefoil, sneezewort, water avens and devil's-bit scabious swayed in the breeze, busy with insects.

For me, the most exciting of all the ecological changes at Carrifran is seeing what's happening to the alpine plants. Halfway up the glen, the Carrifran Burn meanders lazily over a series of gravel bars. River features like these in a typical upland glen are bare, kept open by even the lightest nibbling. At Carrifran, they looked like a little slice of the Alps. Roseroot, sea campion, wood crane's-bill, mountain sorrel and globeflower coloured the gravels, freed from their vertical prisons above. Seeds had been brought down from higher up the glen, bestowed on the rushing water by the survivors on the crags. For the first time in hundreds, perhaps thousands of years, these seeds have the chance to do what their genes programmed them to do: grow into new plants and offer their flowers to the sun.

Many alpine plants are poor competitors; they can only flourish in the absence of more thuggish species. The flow of the burn periodically strips all the vegetation and returns the bars to bare gravel, creating the perfect conditions for alpines to establish. This dynamic process of seeding, colonization, competition and disturbance will mean that the valley will never be quite the same from year to year. This is how nature should be, in a constant state of flux.

Higher up the valley, after a bit of a scramble through the heather and bilberry, pulling ourselves up using the now more scattered and stunted trees, we stopped for lunch near the waterfalls. With some excitement, one of our party spotted some alpine saw-wort, which had narrowly avoided being sat on. This, like many other alpine plants, is a highly palatable species. There is no way this flower could have survived here before the Borders Forest Trust changed the valley's grazing.

Further up the hill, and we found downy willow, which in time will form new patches of montane scrub. Higher still and we found moss-dominated montane heath, complete with dwarf willow, the world's smallest tree, reaching its lofty maximum height of a few centimetres, amid carpets of woolly fringe moss, alpine clubmoss and other hardy, wind-tolerant species.

Alpine Saw Wort

This was the first time I had experienced a full altitudinal succession of habitats like this, from valley bottom woodland to wind-clipped heath. Its recovery here was thanks to an intimate collaboration of people working hand in hand with nature.

The adjacent valley to Carrifran, which acts as a control against which the changes can be measured, is perhaps appropriately called Black Hope. As long as sheep grazing continues in this glen, there is no scope for the ecological changes seen over the watershed to spread. My values tell me that Carrifran is the good half of this pairing, that the restoration of ecological function benefits all of us, makes visits to the valley more enjoyable, and offers hope for the future of wildlife in our hills. But I can't fail to recognize that there is a huge community of people who think in exactly the opposite way, who see Carrifran as a shocking waste of grazing land, who will see the shift from farm to wilderness as eroding traditional rural practices and failing to meet a moral duty to produce food.

In Black Hope, meadow pipits and wheatears, birds of open upland habitat, are maintaining their numbers, while next door in Carrifran they are declining, making way for species like woodcock, blackcap and song thrush as the new woodland takes hold. What Carrifran and Black Hope show together is that there is room in our crowded islands to ensure that there's something for everyone. I would like to see more Carrifrans, but that doesn't mean I want to see the total removal of sheep grazing across the whole of the uplands. With simple measures like fencing, two valleys with sharply contrasting objectives are living alongside each other without conflict and, I hope, with respect.

To compare photos of Carrifran twenty years ago with today is astonishing. The views up and down the valley have been enhanced by the trees. It is every bit as dramatic as it once was, with the craggy peaks of the valley's edges always visible. It is still quintessentially an upland Scottish landscape. The difference is that now there is more birdsong, more insect life, more colour, structure and substance. If I can look at Haweswater by the time I reach Philip's age and see half the changes that he has wrought at Carrifran, I'll consider my efforts worthwhile.

Heading north, the Scottish countryside we drove through seemed even more bleak and empty than it had before. The forestry blocks breaking the open ground were even more trapezoidal and alien, the gloomy grey hillsides even more bare. But we were headed to another bright spot, so our spirits wouldn't stay low for long.

Ben Lawers, which looms over the northern shore of Loch Tay, is the highest mountain in the Southern Highlands. Thanks to the peculiarities of the mountain's geology, which outcrop at just the right altitudes, it provides the perfect conditions for the best suite of alpine plants in the country. A National Trust for Scotland estate that stretches to 4,500 hectares, it is an essential pilgrimage for anyone with even a passing interest in alpine plants or upland conservation in general. The higher reaches of the mountain hide many botanical treasures, but it is the thirty-year-old grazing exclosure along the Edramucky Burn in the reserve's lower reaches that made the most powerful impression on me.

From the car park, a short path across typical grazed hill land leads to a gate in a deer fence. Built in 1990, this fence has given the twenty-three hectares of land within the time to regenerate more or less unaided. Outside the fence, the vegetation is kept at ankle height by the sheep, and skylarks and meadow pipits make up the majority of the wildlife. The inside is like some alternate ecological dimension.

On that first visit in summer, redpoll and twite were flitting around in the birches, drawn in by the abundance of seeds to eat. Every now and then a cuckoo zoomed over the heathery knolls, in search of a hairy caterpillar to guzzle. The preponderance of flowers, richest along the stream sides, included many liberated mountain plants. These, and the shelter of the trees and scrub, provided a living for butterflies and bees, which would have struggled in the windy conditions beyond the fence.

The National Trust for Scotland have set up a family friendly trail which circumnavigates the exclosure, with numbered stops to help visitors appreciate the changes that have occurred over the decades since the fence was put up. Beyond a small amount of planting in the early years, which was soon rendered unnecessary by the accelerating natural regeneration, nature created this oasis all by herself. Given time, and the right starting conditions, she can do so anywhere.

That evening, back in our Fawlty Towers-style hotel in Killin, we chatted about what we'd seen over a few pints. As the sun started to sink, we walked up the river to where we'd been told beavers had recently taken up residence, part of the natural expansion of this hitherto extinct species. Quietly sitting on the bank, we watched as they emerged, transforming from waddling, hairy trolls into sleek and graceful torpedos as they slid into the water. A fitting end to a day full of natural resurgence.

The next morning, with slightly heavy heads, we trekked further up the hill to an even larger and richer exclosure. In

2000 the team at Ben Lawers encircled a 180-hectare lump of mountain called Creag an Lochain with a cleverly designed electric fence, capable of keeping out sheep and deer while withstanding the frequent heavy snow. Rising above the dark surface of the Lochan na Lairige reservoir, which feeds a hydropower station, these crags and the land surrounding them are now home to the most botanically rich habitat any of us had ever experienced in the UK.

The most conspicuous species in the exclosure was downy willow, thanks to the targeted planting of bushes grown from local seed. Rarely growing more than head height, downy willow gets its silvery appearance from a felty covering of hair on its leaves, particularly the underside, which helps to retain warmth and moisture. More established than the few bushes that we'd seen at Carrifran, here they were beginning to join up to form new patches of montane scrub, the rarest of all upland habitats.

Growing through the willows in waist-high stands was a glorious growth of tall wildflowers. Heavy flower heads of melancholy thistle nodded at angelica, wood vetch, water avens and globeflower. Holly fern and alpine saw-wort, two species which grow in tiny patches on our crags at Haweswater, were all over the place. Further up, on low ledges which would have previously been in easy reach of red deer, were the thick leathery leaves of the diminutive net-leaved willow growing alongside frog orchids, indicating the calcium-rich quality of the rocks. We were like pigs in muck, dashing from one species that we'd never seen to the next. Ring ouzels and twite were our constant companions, and up on the windswept ridge we startled a ptarmigan in a mossy heath full of bog bilberry.

The exclosure has been well studied, with plant surveys having been carried out before the fences were installed and periodically afterwards. After eighteen years, there's been a 30 per cent increase in the cover of tall palatable wildflowers, which before the fence could only survive on the ungrazed ledges.

It must be so satisfying for the architects of this exclosure to see that the impacts of removing grazing have been almost exactly as they had anticipated, and it's great that the science has been able to quantify the change – but good God, this is slow progress.

The National Trust for Scotland describes itself as 'the conservation charity that protects and promotes Scotland's natural and cultural heritage for present and future generations to enjoy', and they have owned Ben Lawers for over seventy years. The mountain is a Site of Special Scientific Interest, a Special Area of Conservation, a National Nature Reserve and classified as a category II protected area by the International Union for Conservation of Nature. In other words, it's just about as highly protected as it's possible for land in the UK to be, and all because of its plant life. Yet even here, grazing is still having a negative impact. Only in the exclosures and on the sheer cliffs, which comprise a tiny proportion of the estate's 4,500 hectares, are the plants safe from the sheep and the deer.

This isn't to knock the staff at Ben Lawers, who are all brilliant, knowledgeable and committed people. They are working their socks off to protect and enhance the mountain and its natural inhabitants, keeping the exclosures free of grazers, collecting seeds from fragmentary populations, growing them on and planting them out, pushing the risk of extinction into the long grass for many vulnerable species. But the fact that more of the estate is used for grazing sheep than protecting nature is indicative of the struggle that anyone working in conservation in the uplands has to face. Sheep grazing rights over large parts of the estate are held by third-party farmers, over whom the National Trust for Scotland have minimal influence. Although deer numbers are controlled, large populations on neighbouring land mean that both red and roe contribute to the challenge.

Protecting plants and habitats in the uplands of the UK, even when you know exactly how to do it, is no easy task. Luckily, I was about to get the chance to visit somewhere further afield where nature had more freedom.

CHAPTER 8

Alone in Fidjadalen

FJELLS:
Mountains
(Norwegian)

It's my fifth day in the long glacial valley of Fidjadalen, south-west Norway, the fifth day living out of a rucksack, and the first day when I'm beginning to wonder if I might have been on my own a bit too long.

I've just found yet another incredible swimming hole, its translucent water calm after the churn of the waterfall above. As has become my habit, I deposit my clothes on the warm rock shelf of the bank and totter in, the sweat and dirt of my day's scrambling vanishing into the flow. After a bout of ungainly splashing as I stagger through the stone-slippery shallows, I drift downstream on my back, knowing the chances of another human turning my freedom into shame are close to zero.

I emerge, skin tingling and neurons firing thanks to the cold water. Walking back upstream to my rucksack, barefoot across the soft, thyme-fragrant turf, the image of an adder springs into my thoughts. I haven't seen any snakes so far on this trip, but I assume they must be here, and this looks as suitable a place as any, with warm rocks for basking, and abundant heathy vegetation providing shade and cover for hunting. Getting bitten out here would be bad news; I'm a long way from anywhere.

I wander past my clothes and on up to the waterfall, one of countless many in this glorious valley. The crashing water is captivating, and I allow myself to be hypnotized, river water dripping from my hair. No point getting my microscopic travel towel wet when the sun will dry me so much more agreeably.

Something long and dark tumbles into the foaming pool, a stick perhaps. There are plenty of trees higher up the valley, an incredible number compared to what I'm used to at home. The thing surfaces, swirls around in an eddy and then disappears again under the cascade. I spot it again. Did it just flex? Each time it escapes the turbulence, it seems to swim back in. Slowly, achieving a wider orbit around the boiling water, it begins to move with more purpose, and reveals itself: an adder, obviously stunned by its tumble down the waterfall. Eventually it regains its senses, stops trying to swim upstream, and heads the other way, straight through the pool where I'd been swimming less than five minutes earlier.

In my sun-addled, flower-drunk state, after days with only myself for company, I briefly convince myself that I had summoned the snake. The coincidence of thinking about an adder for the first time since I'd arrived in Norway, and then seeing one minutes later feels auspicious. I'm not remotely superstitious, and tempting as it is to think that I had somehow become spiritually attuned to nature as a result of my solitary immersion in this place, I soon shrug the sensation off, relocate my clothes and get on with my esoteric botanical mission.

The RSPB is an incredible organization to work for. My colleagues are some of the most dedicated and passionate people I have ever had the privilege to meet. The shared zeal for saving species and restoring habitats is palpable; the organization hums with it, and our successes are many and varied. If it wasn't for the campaigning and practical interventions by the RSPB, chances are we wouldn't have avocets, goldfinches, marsh harriers, bitterns, or many other species in anything like the numbers we do today. Sure, it doesn't get everything right all the time, no organization does, but the RSPB fights for the things it believes in and rarely gives up.

A less well-known attribute of the RSPB is that all employees are entitled to a four-week sabbatical for every five years of service, a milestone I reached in 2018. In order to inform our work at Haweswater, I wanted to expand my knowledge of plants, and understand where they fit into a truly healthy, unconstrained upland landscape. This ruled out anywhere in the UK.

On our visit to Carrifran, Andy, one of the group who had shown us around, had recently returned from south-west Norway where he'd taken part in a study week funded through the EU's Erasmus+ professional development programme. He couldn't have been much more enthusiastic about his experiences without it becoming embarrassing. The seed was sown. I'd use my sabbatical to go to Norway and immerse myself in upland habitats infinitely more intact than ours.

After weeks of applying for funding, reading reviews of hiking and camping gear, considering calorific value to weight ratios of various foodstuffs, I had a plan. I secured a place on the same Erasmus+ week that Andy had been on the year before. After that, I had two weeks to explore in more detail, all by myself.

I hadn't been away from Elliot and Aphra for more than a night in their whole lives, and was under no illusions about how much Becki would have to take on while I was away, solo parenting for the best part of a month. I also hadn't ever been away by myself. I'd travelled in my late teens and twenties, but always with others. This was going to be a proper adventure.

No spanners were thrown in the works, funding was received, a new rucksack purchased – the one dating from pre-university travels having been chewed by loft-dwelling wood mice – packed and repacked, itinerary checked and rechecked. Then, after a slightly tearful farewell to Becki and the kids at Glasgow Airport, I took off for Stavanger.

My companions for the next week, a group of Scottish land managers, were every bit as excited about the trip as I was. Our host was Duncan Halley, a native Scot who has lived in Norway for the last twenty-five years and is evangelical about how the UK can learn lessons from Norway to help restore its natural environment. Duncan works for NINA, the Norwegian Institute for Nature Research, and he took us on a deep dive into the history of nature in his adopted homeland. He firmly believes that the only thing stopping the UK from restoring upland habitats to the rich state in which they are found in Norway is politics.

Norway 100 years ago was a very poor country, and the bulk of its population was rural, forced into living hard, subsistence existences. This took its toll on the environment, and swathes of the countryside were completely stripped of trees, so much so that cutting peat for fuel was common practice. No society would choose to burn peat if timber were available. The photographs from this period show a treeless landscape very similar to many upland areas in Scotland or northern England today. Then an accident of history allowed nature to win back some ground: worn down by poverty, hundreds of thousands of people simply upped and left. Mass emigration in the early twentieth century, much of it to the USA, resulted in vast tracts of land being abandoned, and nature was left to its own devices.

If you've spent any time in the Highlands, or the Lake District, much of south-west Norway will feel familiar. Although Norwegian mountains are higher, they are made of essentially the same rocks as ours, sculpted into long valleys by the same period of glaciation. Norway's climate has the same oceanic influences resulting in broadly similar weather, albeit with harsher, snowier winters. It also shares much of its wildlife with us; virtually every plant species I saw in Norway could also be found in the UK, and we share much of their birdlife too. Granted, we're lacking the wolves, lynx, musk ox, reindeer, moose and wolverines from our shores, though some of those are only missing because we made it so.

Steep cliffs in the UK uplands are often adorned with trees. The really inaccessible ones, which are often small and twisted, wouldn't have been worth the effort for someone to climb up and fell them. It is likely that comparable remnants on Norwegian crags are what provided the seed source for the astonishing recovery that occurred once the pressure on the land abated.

A hundred years after mass emigration ended, Norway is one of the richest nations on earth, thanks to North Sea Oil. Its countryside provides a fascinating insight into what can happen when nature has the time and space to operate freely.

The UK enjoyed sublime weather throughout most of late spring and early summer of 2018, and Norway shared in the meteorological glory. For the entire three weeks I was there, I had a day and a half of weather that was slightly less than perfect. As a result, Norway will be forever burned into my memory as a land of eternal sunshine, of crystal-clear fjords, of flowers, birdsong and nights under canvas. It was like I'd walked into a glossy holiday brochure, everything down to the most intimate detail portrayed in vivid, hyper-real colour.

The last week has passed in a blur and, as my new-found friends fly home to Scotland, I wander into Stavanger by myself. While provisioning for the next stage of my trip, I begin to digest a week's worth of lessons from Duncan. With my bag loaded with food, and a head full of ideas, I'm ready to hit the hills and really immerse myself in the landscape.

What I had somehow failed to appreciate in my planning was that Norway basically shuts down on a Sunday. The first bus out of Stavanger is running, but the two more I need to get me to the start of the trail aren't. Duncan had told me that hitch-hiking wasn't really a thing in Norway, and that I wasn't likely to be able to get around that way. But as I don't have much of a choice, I stick out my thumb and plaster on my best smile. I don't have to wait long to be picked up by a university student and his girlfriend (we mainly talk about the metal concert they had just been to). I then score a lift from a Polish office worker, who's been happily settled in Norway for a decade, before a lumberjack in a pick-up takes me the rest of the way. He doesn't speak much English, and with my Norwegian not extending beyond *tusen takk*, we pass the time amiably nodding and smiling at each other.

Waving the lumberjack into the distance at the side of a road in the middle of nowhere, I feel surprisingly calm. Perhaps it is *fjellvant*, the Norwegian word that describes being at ease among mountains. This glacially carved landscape, and the species that call it home, are so familiar to me that I feel far more comfortable than I'd expected before setting out.

The plan for my next six days is a simple one. I'll walk from the roadside up to the head of the valley of Fidjadalen and then follow the river from its source to the great waterfalls at Månafossen, where I can pick up the road again and travel back to Stavanger. I had visited Månafossen with the Erasmus+ group earlier in the week, so I know where I'll end up. Along the way I'll study the vegetation, hoping to gain an insight into how it all fits together. I need to understand where the plants I know from home like to grow when given the opportunity to choose for themselves.

I'd chosen Fidjadalen not because it was somewhere known for being particularly botanically rich; indeed, by Norwegian standards, it is considered pretty dull. I chose it because it is about as comparable to the Lake District as a place can be. Although it's on roughly the same latitude as Orkney, Fidjadalen has acid rocks, steep terrain, high rainfall, a short growing season and broadly comparable summer temperatures, all of which makes it about as similar to home as I can get. Haweswater and Fidjadalen are about 460 miles apart, roughly the same distance that separates Inverness and London.

⮜⮞

I plan to spend my first night at Sandvatn, a little over 900 metres above sea level. The first part of the trail from the road has some close-packed contours, which, combined with the weight of my pack, result in not a little perspiration. So, once I crest the lip of the valley, I'm pleased to see a landscape pocked with water. Swimming had become a compulsion during the previous week. Fuelled by the collective enthusiasm of all the members of our Erasmus+ group, no swimmable river, fjord, *vatn* (lake) or *tjern* (tarn) had gone unswum. I fully intend to continue this newly established tradition, and to break the group record for highest-altitude swim, which currently stands at a shade over 1,100 metres. The weather is perfect and there's nothing on my agenda for the next six days apart from walking, swimming and flowers. I've never been so free.

The Norwegian Hiking Association (DNT) maintains 22,000 kilometres of trails, marked by red-painted T-marks. Thanks to these, it's hard to get lost, but because the paths aren't surfaced, using them does nothing to detract from the sense of being in the wild. The Ts allow me to focus my attention on understanding the nature of the place, without having to waste concentration on navigating. The DNT also look after 550 hiking cabins. I wasn't planning on using them, which as it turns out, is a big mistake.

The term hiking cabin for me conjures something cramped and basic, where I'd be forced into close proximity with others – not something I want for this part of the trip. It turns out those concerns are totally unfounded. All the DNT cabins that I come across are spacious, clean and comfortable, and constructed in sympathy with their pristine surroundings. The cabins all come in pairs, the theory being that if one burns down, there will be a spare. Stocked with firewood, log burners and gas for cooking, they all have fully equipped kitchens, beds with mattresses, duvets and pillows, composting loos and bags of Nordic charm. Some are even supplied with food to buy. The whole system runs on trust, and access to the cabins is gained by a universal key, which DNT members get when they join. For the unmanned cabins, there is no need to book;

they're crammed with so many beds that a space is guaranteed, even at busy times. You help yourself to a bed, collect water from the nearest lake or river, and then drop a form with your credit-card details into a slotted safe, declaring what you owe on departure. Users are expected to clean before they leave.

If it sounds like I'm advertising, then perhaps I am. These cabins are amazing, and I really should be making better use of them. As well as being a valuable amenity, they are also testament to the Norwegian social spirit. The trails and cabins are maintained by local associations, and the majority of the work that goes into looking after them is done by volunteers, who collectively put in over 750,000 hours a year.

My slightly flawed plan is to camp near to the cabins, paying as a day guest to give me access to the facilities, but without sleeping in them. I figure this would give me a sense of security while avoiding the need to be cooped up with anyone. I really shouldn't have worried. For the whole week, I only see one other person in the vicinity of the cabins. If I'd known how good they were, and how quiet, I could be travelling with just a pillowcase and a sheet. Instead, I've loaded 10 kilos of tent, sleeping bag, roll mat, cooking gear and other camping paraphernalia onto my protesting shoulders. I'll know for next time.

The layby where the lumberjack dropped me was obviously also an unloading point for sheep, and this first stretch of the walk is about the first place I'd seen so far on the trip that is noticeably impacted by grazing. It looked as if sheep were herded up through the same small pass I'd just followed, before dispersing. I've arrived ahead of them, but a couple of discarded ear-tags from the previous year, and the relative frequency of coarse grasses, are testament to their presence.

The landscape on this first and highest part of my route is open and windswept, with bare rock making up the lion's share of the broad plateau. It is the first week in June, but there are still lingering snow patches and ice in the deeper tarns. The vegetation that does manage to grow, where it isn't covered by snow, grows in patches wherever there is enough soil to sustain it. Closest to the road and the sheep pens, these patches are grass dominated, but as I push further on, they begin to get much more interesting.

Only the toughest of plants can survive in a habitat like this, ones which have adopted a prostrate growth form, able to tolerate extremes of moisture, desiccation, wind, heat and cold. These are true montane heaths, made up of trailing azalea, alpine clubmoss, bog bilberry, cloudberry, alpine bearberry and dwarf willow, growing in a matrix of woolly fringe moss. All tiny, stunted by the conditions, but virtually all offering flowers and berries to feed pollinating insects and birds. There are mountain ringlet butterflies up here, and chunky hawker dragonflies drone past. Alongside the meadow pipits and wheatears, ptarmigan are the most noticeable bird, launching into their panicked flight only when I'm almost on top of them, their dumpy white bellies skimming over the plateau.

Many of the exposed rocky tops of the Lake District hills, including Harter Fell, have similar climatic conditions to this Norwegian plateau, but the montane heaths that used to clothe them have been dramatically altered by sheep grazing, and in an alarmingly short space of time. Derek Ratcliffe noted that the loss of this most sensitive of habitats was the most complete and conspicuous vegetational change on the fells in his lifetime. Although we probably didn't have trailing azalea or alpine bearberry in the Lakes, bog bilberry still just about manages to cling on, as does cloudberry, dwarf willow and alpine clubmoss, but rarely in places that the sheep can reach them. Even the woolly fringe moss, which according to Ratcliffe used to dominate swathes of fell top habitat, is now drastically reduced, in many places restricted to inaccessible boulder clefts, ledges and wall tops. The sheep don't even eat it, but their trampling, dunging and the browsing of the plants that grow through the moss are enough to have knocked it back, allowing coarse grasses and rushes to dominate instead.

I reach the Sandvatn cabins by early evening, pitch my tent a respectful distance away in a cosy moss-lined depression between the rocks, and settle in for the night. Although occasionally punctuated by the eerie sound of cracking and colliding ice sheets in the nearby tarn of Sandvatnet, my sleep is long and peaceful.

Seeing this missing piece of the Lake District jigsaw on the tops of these Norwegian hills is fascinating, but there are only a handful of plants that can cope up here, so after my 'just add water' breakfast, I pick up the trail again and head down the valley for richer pickings.

For the first couple of miles I'm still on the plateau, where small tarns and bare rock are interspersed with patches of dwarf heath. As I begin a slight descent, I notice a shocking tuft of pink growing out of a small cleft by the edge of the path – alpine catchfly! At home, this a plant of exceptional rarity, found in only two locations in the whole of the UK. The larger of the two British populations, on a remote hilltop in Angus in the Highlands, holds tens of thousands of plants. The second, far smaller one is in the Lake District's Western Fells, growing in a single steep and crumbly gully. Alfred Wainwright, father of Lake District fell walking, had this to say about the location: 'In shadow, the scene is sombre and

Alpine Catchfly

forbidding. The silence is interrupted only by the croaking of the resident ravens and the occasional thud of a falling botanist. This is a place to look at and leave alone.'

The explanation for alpine catchfly's strikingly binary distribution in the UK is thought to be mineralogical. It can tolerate much higher concentrations of heavy metals in the soil, particularly of copper, giving it a competitive edge over other species that can't cope. This heavy-metal association is neatly made by the proximity of the plant's Lake District locality to the region's last working metal mine. Historically, alpine catchfly had been used by mine prospectors, its presence indicating ore deposits near the surface.

Needless to say, I'm excited to see it, and quickly begin recording its grid reference and bending myself in a range of contorted poses to try to capture the best photo. Then I notice another plant a few feet away, surrounded by a low growth of grey lichen. Then I see another, and another. In an area the size of my bedroom, there are probably more alpine catchfly flowers than in the whole of England. Over the course of the next few miles, I keep coming across patches of them. Clearly this isn't a notable species in Norway. Perhaps it used to be more wide-spread in the UK too; there's no real way of knowing.

As I descend from the rocky plateau, the path leads along the side of a deep trough, a pool at its base. This is the first place along my route so far where the terrain offers some shelter from the wind, and immediately a different suite of plants appears. A juniper, clinging to a rock above the pool, is the highest tree I find in the valley, growing at about 980 metres, roughly the same altitude as the top of Scafell Pike, England's highest peak.

This prompts a revelation. Up until this point, I'd always thought of the tree line – the point above which trees are unable to grow – as being set by altitude. In the highest mountain environments, like the Alps, this is indeed the case, but in lower-lying ranges, like the fjells of southern Norway and the Lake District fells, it's exposure, not altitude, that sets the tree line. The desiccating effect of the wind blowing across the mountaintops means that only squat and leathery-leaved species can survive. As soon as the wind drops, the trees can grow.

Rounding a rock at the end of the trough, the transition to sheltered conditions is manifest at scale, and a lush wooded landscape is spread out before me. Below the windbreak of the valley sides, millions of trees are thriving. Twisted into an endless variety of contortions, a specially adapted high-altitude form of downy birch, known as mountain birch, is the dominant species along the tree line. Scattered throughout the heathy landscape, these mountain birches can withstand months of winter snow cover and avalanche. Most are shorter than me, some are prostrate, some leaning nonchalantly against boulders, all are weather-sculpted knotty testaments to survival and rebirth – a century ago, this whole landscape was open and treeless.

Genetically, birches are a bit of a muddle. The two commonest species in the UK, downy and silver birch, look very much alike and can hybridize with each other to form a spectrum of intermediate forms. Downy birch tends to occur more in the north and at higher altitudes, silver birch more in the lowlands and in the south. A third species, dwarf birch, which rarely grows more than waist height and has lovely round leaves with wavy edges, is much rarer than its two larger relatives. It's thought that mountain birch originated from the repeated crossing of dwarf and downy birch over the course of millennia. Whatever their origins, their tortured forms are a vital part of this Norwegian landscape. In the UK, dwarf birch only occurs in the Highlands, bar a scattering of tiny populations in northern England and the Borders, and mountain birch only survives in isolated refuges on cliffs and crags. There's no doubt that both were once far more widespread.

Hazy, silvery-green patches of downy willow adorn the sloping valley sides. This is the montane scrub habitat I was longing to get to know. The squat, open growth form of these willows allows plenty of light to reach the ground below, and here is the flower-rich plant community I know from the crags on Harter Fell, with roseroot and wood crane's-bill, beech fern and great woodrush, starry saxifrage and alpine lady's-mantle. Our 'tall herb ledge vegetation' isn't a plant community as it should be – it's montane scrub that's missing its trees.

Between the birch and willow scrub are large treeless areas. Many of these are open because their peaty soils are too wet for trees to grow, dominated instead by colourful sphagnum mosses and studded with carnivorous sundews. Species-rich heath grows in rocky areas where the soil is too thin for tree roots to gain purchase. Patches of bare rock support lichens and mosses. The river and innumerable pools provide still more diversity.

A couple of miles into this wooded wonderland I reach

Blåfjellenden, which is to be my base for the next two nights. I pitch my tent a short way from the paired wooden DNT cabins which sit on a rocky bluff with a commanding view. Down the slope, two of the valley's rivers meet in a wide pool. The water is impossibly clear; it is like swimming in gin, perhaps more intoxicating, the trout in full view metres below. Rivers are a powerful indicator of the health of the land through which they flow. For a river to run this clear, through a landscape with as much peat as any part of the UK uplands, shows just how intact this landscape is.

I haven't seen anyone on my walk from Sandvatn, and nobody is using the cabins at Blåfjellenden. All the doors are open, so I take a look around, and pay the day rate again to use the facilities. Leaning up against the outside wall of one of the cabins, a half-whittled rowan stick, five feet long and as thick as my wrist, has been left by someone. Stripped of its bark, and partially shaped with a penknife, it has a simple natural grace. Over the course of the next few days, the stick becomes my prized possession, a protective talisman in my short spell of solitude.

I want to digest as much of the detail of this place as I can. Using a similar method to the one that we use for surveying plants at Haweswater, I walk a series of transects that start at the valley bottom, stopping every 50 metres of altitude to record the species growing along the length of my rowan stick. As well as the ground flora, I make notes on the trees at each stop, and how they change with altitude and exposure. I record the bird life, the soil conditions and take photos at each point to refer back to.

The two days I spend exploring the area around Blåfjellenden, drenched in sun and cut off from the world, are about the happiest I can remember. For years to come, on the nights that I struggle to sleep, my thoughts will be calmed by the river pool in its landscape of flowers.

CHAPTER 9

The Summer Farm

BLÅFJELLENDEN:
The end of the blue mountain
(Norwegian)

My stick-assisted surveys in Fidjadalen are as much about giving me a focus as generating hard facts and figures. They ensure that I really look at the landscape, its habitats and species, rather than just drifting through it. They force me to take routes that I might not have bothered with, to notice details that I might have otherwise missed.

Many of the flowers that I come across are exceptionally rare in the Lakes, but it's also striking to see some which, using my UK frame of reference, are growing in the 'wrong' places. Bog rosemary, bog bilberry, cloudberry and dwarf birch are all considered bog species at home, but here in Fidjadalen they are far less choosy, and often grow larger away from the nutrient-poor, waterlogged peat bodies.

Bogs offer a degree of protection from grazing. Most animals avoid them, as there isn't generally much worth eating to justify the risks of getting stuck or the irritation of the abundant insect life. Over time, the impact of heavy grazing has forced several species into refuges in bogs, in the same way that it's pushed others into the crags, by nibbling them into oblivion in the drier and more accessible places. That's affected the way that we think of them, and even how we've named them, but it's just another part of our land-management legacy.

Tormentil is the commonest wildflower on the sheep-grazed Lake District fells. Its small yellow flowers have four petals, so my kids call it four-mentil. Its official, scientific name is *Potentilla erecta*, yet when you see it growing in the cropped pastures of the Lake District fells there's nothing erect about it. On the Plantlife webpage for the species, they describe it as 'a creeping, prostrate plant'. I'm a bit confused when I see tormentil for the first time in Fidjadalen, as here it lives up to its Latin description, standing proud in a thick sward of bilberry and beech fern.

Carl Linnaeus, a Swedish botanist born in 1707, is considered the father of modern taxonomy. He formalized the binomial system of naming species that's still used today. Common names for species, be they plants or animals, vary from country to country, and even from region to region. The Latin name, however, is universal. The first of the two words in a

Tormentil

species' Latin name is its genus, indicating the group of closely related species to which it belongs. We are the only extant member of the *Homo* genus, all our closest relatives such as *Homo neanderthalensis* having died out thousands of years ago. The second part, our *sapiens*, belongs to our species only. The Latin name often gives a useful hint about species' characteristics, or habitat preference, and so having a basic understanding of Latin is quite useful as a naturalist.

As well as coming up with the system, Linnaeus was also the first person to apply it to a multitude of species, officially describing them in his new rigorous scientific context. The (L.) that appears after a Latin name in many botanical textbooks indicates that the species was described by Linnaeus. *Potentilla erecta* (L.), or tormentil, is one of his, and, being Scandinavian, I guess he would have been used to seeing it growing erect, as it does in Fidjadalen. By contrast, in the grazed pastures of the English uplands, tormentil has adapted in order to survive, hunkering down at ground level so that the teeth of grazing animals can pass overhead. I find it astounding that the history of how we have managed our upland landscapes is reflected in the posture of this humble yellow flower.

My biggest non-botanical eye-opener from Fidjadalen is to do with ring ouzels, which have the rather unfortunate Latin name of *Turdus torquatus*. The RSPB website says that ring ouzels breed in 'steep-sided valleys, crags and gullies' and that 'nests are located on or close to the ground in vegetation (typically in heather), in a crevice, or rarely in a tree'. Fifteen or so pairs nest each year at Haweswater, way up in the crags and corries, including in Sale Pot where I'd surveyed for them a month ago. At home, ring ouzels are thinly spread, sometimes heard and rarely seen. When I arrive at Blåfjellenden, the ring ouzels are chasing each other around the cabins, squabbling every bit as vigorously as their close relative, the blackbird.

Over the course of a few days, in a landscape of birch and willow scrub, I see and hear more ring ouzels than I've ever seen in my life. At home, ring ouzels tend to stop singing at dawn, but here in Fidjadalen, they seem to be doing so almost constantly, particularly in the highest of the wooded areas near the tree line. I watch recently fledged chicks perch on birch branches, getting fed by harried parents. I record them on many of my survey stops, singing alongside willow warblers and chaffinches, not species that I'd ordinarily put together.

The ouzels are more vocal because they are at a higher density here than at home, so the competition for territories and mating rights is more intense. They're here at higher density because the habitat is more to their liking. Ring ouzels have a similar diet to blackbirds, pulling up worms from the soil, catching insects and eating berries and fruit when they're in season. The berries are particularly important, as they provide the fuel the ouzels need for their long autumn migration to North Africa. They're popular with the local human population too – berry picking is a traditional autumn activity in the Nordic countries. With such an abundance of juniper, rowan, bilberry, cloudberry, cranberry, cowberry and crowberry, there's plenty to go around for the ouzels and the people.

Ring ouzels don't want to be a species of burned moorland and heavily grazed pasture, it's just that the crags that are dotted around these landscapes in the UK harbour the closest thing they can find to montane scrub, the habitat to which they really belong.

As I'm packing up my tent after two nights at Blåfjellenden, I notice a woman and a boy emerging from the *seter* on the other side of the valley. *Seters* are summer farms, which historically were where farmers from lower elevations would move up to for the summer months with their sheep and cattle. I suspect that the ruins at High and Low Loup may have once been used in a similar way. Most *seters* were abandoned during Norway's period of mass emigration, but some now are coming back into use again. I'm keen to find out more about how the farming up here works, so I jog across the valley to catch up with the woman and the boy, who I learn is her son. Quite what they think of me, my rowan staff and my total lack of Norwegian is anybody's guess, but they're happy for me to walk along with them for a while and to pick their brains.

Magda is in her early thirties, with an athletic build and perfect English. They're heading down to the road to catch the bus back into town in time for Magda's son to go to school, having both spent the last couple of days working with her father to herd his sheep up into the valley for the summer's grazing. I didn't see the flock arriving the day before, having been lower down the valley, flower counting.

She tells me that her father will be staying up in the *seter* for the next few months, keeping an eye on his flock of ninety-three sheep as they graze the valley's woody heaths and pastures. Now, in June, is the traditional time for sheep to be turned out to the hill, as soon after the snow has melted as possible. They'll be up here until September or October. It would be a disaster to leave it too late and for the flock to be buried in the snow.

Magda's father's flock had been unloaded at one of the roadside pens I'd passed in the lumberjack's truck. Magda and her son, her parents, along with their neighbours, who were also bringing in their flock, had walked the sheep the six miles up to the *seter* with the help of a few well-trained dogs. She describes how well the sheep know the valley and the route into it, and how each flock has its own territory – a system that sounds identical to hefting in the Lake District. I try to calculate from her totting up of all the different *seter* flocks how many sheep would be in the valley once they'd all arrived. Dividing the number by a crude estimate of the valley's extent, it works out at roughly one ewe for every five hectares.

One of the things that makes sheep farming in the Lake District such a challenge is disease. Ticks and intestinal worms are a serious problem, and most farmers will gather in their sheep to medicate them several times during a summer grazing season to keep these parasites and the diseases that they spread at bay. I asked Magda how this compared with the care of her father's flock. She describes the fjells as being like a spa for the sheep. They need nothing from their owners while they're up in the hills, coming down healthier than when they go up.

It occurs to me that I haven't seen a tick since being in Norway. I've been walking through plenty of places that I'd have considered very 'ticky' at home but haven't picked up a single blood sucker. Because there's such an abundance of food on offer, there's no need for the sheep to go into wet ground which might harbour worms. Magda tells me that each year they usually lose a few sheep to wolverines (voracious predators in the badger family) – or just lose them – but that this is an accepted part of the system.

Magda's father is a full-time farmer. His home farm near Sandnes, close to Stavanger, is an hour or so west of the roadside unloading pens. He used to have a dairy farm, but a battle with cancer had prompted a change to farming with a more manageable workload, and now he just keeps sheep. As in the Lakes, it's government grants more than livestock income that sustain his farm business. He receives a grant for grazing on the hill in order to try to prevent the birch from taking over entirely, and he gets another for carrying out small-scale burning. This helps to keep the birch in check and promotes the growth of young heather, to boost the number of grouse which have a liking for the fresh shoots. I'd noticed some of his burned patches as I'd approached Blåfjellenden, which had seemed incongruous in such an otherwise wild landscape. They were generally pretty small, 10 metres or so in diameter – a very different sort of heather burning to the industrial-scale rectilinear blocks that are the hallmark of the intensive grouse moors of the British uplands.

Magda and her family take justifiable pride in what they are doing. She talks a lot about bringing traditions back to life and reconnecting family ties to the land. It's a real eye-opener for me, to see how a farming system with sheep could work hand in glove with a rich natural upland environment.

After half an hour of amiable chat, I thank Magda for sharing her knowledge and experience, and head back to the cabins to finish up my packing and to have a quick last bath in the river pool.

For the rest of the day, I follow the river as it winds its way westwards down the valley, in and out of lakes like a string of pearls. There is endless variety in its character. Where the valley is broad, it divides into braiding channels, rippling across gleaming gravels, heaven for spawning fish and flown over by common sandpiper and dippers, Norway's national bird. With steeper gradients, it becomes a torrent again, rushing across bedrock and boulders. Where there is a step in the geology, it transmutes into a pounding cascade, tumbling into deep, foaming cauldrons. Where the river flows through rocky gorges, my instincts compel me to peer into inaccessible nooks and crannies for plants that might have taken refuge. These are the places that often yield surprises at home, but I don't find anything on the ledges that I haven't already seen growing in the wider landscape.

At one point the trail crosses a crazed pile of boulders, the legacy of an ancient rock fall which had blocked the entire width of the valley, damming the river and backing water up into a lake called Månavatnet. The scar marking where the rocks had parted company with the valley side can still be seen. In order to continue its journey downstream, Månavatnet's water must drain through this gigantic boulder field, making it act as a natural store for flood water; the boulders throttle the flow of water, meaning that the lake above rises disproportionately compared to one with a clear opening.

The Hiking Association's trail makers have done an astounding job of marking the route through this rocky jumble. Jumping from car-sized boulder to boulder, following the red Ts, I have faith that the route won't result in an ankle-breaking plummet, a labyrinthine dead end or a disappearance into a subterranean river. Looking down I find interrupted clubmoss, another species which only grows in one area in the Lakes and nowhere else in England, nestled in mossy patches between the boulders.

As I make my steady progress down the valley, its habitats continue to develop. At the highest elevations birch and willow were the dominant trees, with scattered juniper and rowan. As I descend, the species mix gets more diverse. I start seeing aspen at around 600 metres above sea level, bird cherry at 550 metres and hazel at 400 metres. The trees tend to be more closely spaced now, forming proper woodland, rather than the open, scrubby medley that I'd been walking through higher up. The ground flora remains flower-rich throughout, with melancholy thistle and wood crane's-bill growing alongside the bilberry and dwarf cornel. I come across some twinflower, a delicate species with paired drooping pink flowers that's restricted to Scotland's fragmented Caledonian pine forests in the UK. It's enthralling seeing where all these species slot into the overall mosaic, having had decades of freedom to find the places that suit them best.

⚶

I spend my last three nights camped by the river not far from Mån farmhouse. Last used for farming in 1915, the building, complete with a turf roof out of which young birch trees are growing, has been sensitively restored as group accommodation. The lower section, which would likely have been used for storage of livestock or hay, is now a small museum, telling Fidjadalen's history.

My remaining days in the valley fly past, as I survey its steep sides, catch fish and swim. On my final transect, I come across a herd of ten highland cattle, grazing the rocky pastures above the farmhouse. Other than Magda's sheep flock at the opposite end of the valley, these are the first grazing animals I've seen. They're wearing cowbells, an essential aid to finding them in this big complex landscape.

It's easy to tell roughly where the cattle have been from their dung, but there are clues in the vegetation too. The area in which they roam still has plenty of trees, but they are a bit more widely spaced now, and the ground flora seems more diverse. There's a greater variety of willow species in the sward, including creeping willow, which is woven in with the ferns and flowers. I see bearberry and dwarf birch for the first time since arriving in the valley, as well as more orchids than I've seen elsewhere. As I study these rich pastures, I realize that there's one species I'd expect to see lots of at home in a place like this, but that I've hardly encountered at all.

Bracken, the world's most widespread fern, probably isn't on anyone's list of favourite plants – it's certainly not on mine. Walking through it in the summer when it can grow to over two metres high is a nightmare. The thick stems seem to grab at the ankles, tripping you with every step. It harbours ticks, and a day of bracken bashing necessitates a thorough check in case any of these tiny parasites have found their way into lesser explored bodily regions. This was a horror I didn't have to endure in Fidjadalen.

Bracken is a big issue in the UK. It's at its most problematic in the hills and is considered dominant across more than 4 per cent of the entire upland area. At its most dense, bracken casts almost complete shade, so that few other plants can grow below it. Sheep can easily get lost inside a bracken bed, and be missed when the rest of the flock is gathered in. It's a major problem for farming, for nature and for access. It does, however, provide more clues as to the history of the Lake District landscape.

In a pristine environment, bracken is a fern of woodland glades and edges, requiring lots of light. Thousands of years ago, when the hills of the Lake District were as wooded as those I was walking through in Norway, bracken would not have been common. Some 5,000 years ago, during the Neolithic period, our ancestors were starting to become more settled, felling the trees with stone tools to fuel fires and build with. As the glades and clearings increased in size, eventually joining up to create a more open landscape, the bracken was able to spread.

Until relatively recently, bracken was widely harvested, which kept it in check to a degree. It was used as thatch, as livestock bedding, as a mulch for growing turnips and potatoes, as packing material and fuel. The fresh green tips, known as croziers or fiddleheads, are edible, and I'm told taste like asparagus, with hints of almond and kale. As delicious as that sounds, without careful preparation bracken is carcinogenic, which slightly dulls my enthusiasm for trying it. Even if Hugh Fearnley-Whittingstall got behind bracken as the latest super-food, I doubt it would do much to knock it back on the fells.

Nobody harvests bracken in the Lake District any more, or at least almost nobody. For the past few years, Simon and Jane, who farm up the road from us at Haweswater, have been cutting and baling it from the surrounding fells to create a novel product. After years of trialling different recipes, they've perfected a compost comprising bracken, which is rich in potassium, and sheep's wool, which helps to retain water. They've made a successful business turning two products that have almost no value into an environmentally friendly, peat-free growing medium. We buy their compost in bulk for use in our tree nursery, where we grow juniper, hawthorn, wych elm, oak and other broadleaf trees, as well as mountain flowers, all of which we plant back out onto the fells. A nicely closed loop, and one which helps to keep money circulating in the local community.

Bracken used to be kept under control partly by the trampling of large grazing animals like cattle and ponies. In Naddle

Valley, where we have a small herd of hardy belted Galloway cattle, the bracken is visibly weaker on their side of the fence compared with the side that is only grazed with sheep. The increase in bracken is often blamed on the slight reduction in sheep numbers that's occurred since the beginning of this century. Sheep don't eat bracken, and they're not heavy enough for their trampling to have an impact on it either. In a fenced area of woodland planting on Bampton Common that was installed in 2011, the bracken is just as thick on the inside where there are no sheep, as on the outside where the sheep still graze. If sheep kept the bracken down, you'd expect to see at least some difference after a decade.

Bracken only thrives where the soils are deep and dry enough for trees to grow. You don't find bracken in bogs, or on the exposed fell tops. Bracken therefore indicates the historical extent of woodland on the fells. Farmers hate bracken, walkers hate bracken, sheep don't eat or trample it and it crowds out other plants. In the steep slopes where grazing it with cattle and ponies isn't practical, the obvious thing is to plant trees into it. Once under the shade of a tree canopy, bracken quickly weakens. I find superimposing trees onto hillsides dominated by bracken a useful way to reimagine the Lake District.

In Fidjadalen, no reimagining is needed. In a land of wooded fjellsides, I see barely a single frond of bracken. I don't miss it.

At the top of my final transect, I watch the cattle confidently scale some steep slopes, totally at home. Way below them on the distant valley floor, I can make out a large steel-framed farm building, perhaps a shelter for these beasts during the winter months. It seems that the disturbance caused by the action of the cattle is a plus for biodiversity, as well as providing a local farmer with some income.

Grazing is a key component of virtually all the world's ecosystems, helping to maintain a balance of open and wooded habitats. I've not seen a single deer track while I've been in the valley. I know there are red and roe deer, as well as the much larger moose, or European elk, in the area, based on the information that Duncan had given us, but they're obviously here in very low numbers. I'd seen signs of moose, on the Eramus+ week, and there was no mistaking their dung or the tracks they made through the thick vegetation, but I've not had a glimpse of them in Fidjadalen.

During Norway's period of rural poverty before mass emigration in the 1800s, its deer had been over-exploited to keep starvation at bay. By 1900 roe deer were extinct, red deer numbered in the low hundreds and moose were only present in the east, along the Swedish border, where hunting pressure was lower. Since then, numbers of all three species have increased dramatically. Recent estimates suggest there are a little over 100,000 each of red deer, roe deer and moose in Norway, and 40,000 or so wild reindeer. In the UK, which is only two-thirds the size of Norway, there are thought to be more than 1.5 million deer. That equates to almost seven times the density – no wonder we don't have as many trees.

Recognizing the risks that large deer numbers can pose to road safety, to forestry, as well as to wildlife habitats, Norway now operates a carefully regulated system of hunting to keep its deer populations in check. Permits to hunt a specified number and type of animal are provided to landowners, who can trade them to generate income. Hunters must have a licence to shoot deer and demonstrate their proficiency by taking an annual shooting test. Kill returns from the hunters allow the government to adjust how many licences they issue, ensuring that populations are kept at a sensible level. Because the numbers are low, hunting deer is a skilled and patient business, as much about the chase as it is about the kill. It's the norm for the hunters to take what they kill home to eat.

Deer management in the UK is a very different business. Where it happens at all it's often poorly coordinated and dominated by large landowners with a vested interest in keeping numbers high to generate income through paid stalking. Shooting rights are often retained by landowners, preventing a tenant farmer from doing anything about the deer, even if they are eating crops, damaging trees or knocking over walls. Deer are a big challenge for us at Haweswater, and their browsing has a serious impact on the trees we plant and those that should be naturally regenerating. We have both red and roe deer to contend with, and while their numbers are not at the same levels as in some parts of Scotland, there are still too many of them as far as restoring woodland and scrub is concerned. Like everything we do at Haweswater, this is an issue we can't solve by ourselves, so we are working with United Utilities and the local deer-management group to cull the deer and bring their numbers down to a level that will allow the landscape to recover.

Norway's small deer populations combined with the cessation of farming allowed a century of virtually unchecked habitat recovery across much of the Norwegian countryside. Tiny pockets of woodland and scrub on steep crags supplied a source of seed that rained down on the land below, enabling woody habitats to reassert themselves.

This regeneration isn't universally popular. Because of its dominance, birch is considered a weed by many, and continuing woodland expansion is squeezing species that are adapted to more open habitats. The grants Magda's father receives for grazing and burning are part of wider attempts to try to keep more of the landscape open. As farming returns to Fidjadalen perhaps the slow pendulum of habitat change might start to swing in the opposite direction.

My last full day in Fidjadalen is a Saturday. As I descend from my final survey transect, I see that the old meadows around the farmhouse have sprouted tents, like gigantic polyester toadstools. As this is the end of the valley closest to the road and the car park for the Månafossen falls, weekenders have arrived to spend time in the great outdoors. One tent has appeared close to my camp, and three twenty-somethings have waded out onto a rock in the middle of the river to smoke. As I get chatting with them, they tell me that due to the prolonged dry spell and the resulting risk of wildfire, there is a nationwide ban on naked flames. It seems I've been breaking the law with my gas stove every breakfast and dinnertime, but then I haven't really been following the Norwegian news much. The middle of the river is the only place my new neighbours feel they can smoke safely.

They're all professionals, a policeman and two working in IT. They spend most of their weekends out in the hills, hunting, hiking, fishing and skiing. Like everyone else I've met, they were polite and friendly, and perhaps a bit baffled why anyone would travel from the UK to look at flowers.

It turns out they're heading back to Stavanger the next morning and offer me a ride. After a dinner comprising the last of my supplies that don't need cooking, I drop into sleep listening to the sound of the river and the ring ouzels chacking in the birch trees.

As the plane begins its descent into Manchester, and the familiar cluttered British countryside comes into view, I worry that my perception of it might have altered. Would it seem depauperate compared to all that Norwegian wildness? The bustle of the airport and baggage claim push the thoughts from my mind, and it's only when I'm on the train that I start to look at the countryside through new eyes.

Certainly, the hills in the Forest of Bowland that I can see through the train window as I travel north out of Preston look even more scalped and barren than they'd seemed before, but there are positives too. Without realizing it, I'd missed the aged craggy trees and the thick hedges that are such a feature of our countryside. Norway's recent recovery from an almost treeless state meant that I'd hardly seen a tree over a hundred years old. The landscape's self-willed recovery there was still in its early stages, and the pioneer trees, the birches and the willows, dominated. The countryside I'm passing through seems more mature, with proud old oak and ash, sycamores and cherries. As the train enters Cumbria, I can pick out hazy yellow hay meadows reaching their flowery peaks. Although the UK's ancient hay meadows are now vastly reduced, Norway's have been lost almost entirely.

A year later, I get the chance to learn a bit more about Norwegian hill farming, thanks to an event organized by the Foundation for

Common Land at the University of Cumbria's picturesque campus in Ambleside. They'd invited a group of Norwegian farmers over to the Lakes for some knowledge transfer. I'd picked up bits and pieces about how farming worked from my conversation with Magda and presentations from Duncan, but these had focused mainly on the care of livestock while they were up on the hill, and I couldn't really picture how the whole system worked.

Norway seems to do a lot of things right. Taxes are high, but so are wages, and everyone receives a good standard of free education and healthcare. Despite being an economy based on oil, it has many solidly green policies. Most of its energy comes from renewable sources, mainly hydro, and electric cars are subsidized and common. Governance tends to be local and inclusive. Forest schools are the norm for pre-school age children, engendering a deep connection with the outdoors from an early age. Hiking, skiing and hunting are popular leisure activities with a well-managed and sympathetically built supporting infrastructure. Like much else in Norwegian society, farming is also run in a thoroughly efficient way.

Because such a large proportion of Norway is mountain, hill farming is the norm, and the system is in many ways comparable with how it's practised in the Lake District. As in the Lakes, Norwegian hill farming is adapted to the local climate. Livestock are moved up onto high-altitude pastures during the spring and summer months, allowing lower-lying fields to grow into hay or silage, which is harvested to feed the animals through the winter. As in the Lakes, it is a system dominated by sheep, and one that is heavily dependent on public funding. Norwegian mountain pastures aren't usually owned by the farmers who hold rights to graze them; the same is true of most Lake District common land. Neither the fells nor the

fjells tend to be fenced, and sheep are hefted to their own traditional areas. Gathering is carried out communally.

There are some fundamental differences too. The most
prominent of these, at least as far as the impact on the environment goes, is numbers. The Norwegian farming delegation at
the Ambleside campus confirmed that sheep in their summer
pastures typically graze at an average density of about one ewe
per five hectares, for four months or so a year. Clearly there's
plenty of variation around this figure, but the farmers agreed
that this was about right. There's lots of variation in the Lake
District Fells too, but the average comparable figure is roughly
one ewe per hectare, five times the density in Norway, and
sheep typically graze our fells for a much longer period.

In the UK, the average sheep flock comprises 422 ewes – the
largest in Europe. In the Lakes, it's higher than the national
average, at around 650. In Norway the average is just seventy-
three. Because Norway experiences longer periods of snow
cover than the UK, its farmers have little choice but to keep
their sheep inside heated sheds between October and April.
The ewes are put to the ram indoors, and they stay there until
giving birth five months later.

Because the growing season is short, Norwegian farmers
throw everything they can at their valley bottom pastures. All
the muck accumulated in the shed over the winter months gets
spread on the land, and synthetic fertilizers are often added to
try to boost growth further. Norway applies nearly twice the
amount of nitrogen and phosphorus fertilizer per hectare of
farmland as the UK, an intensity which has erased virtually all
their traditional hay meadows, which were as much a feature
of their farming before the advent of fertilizers as ours. However, with only 3 per cent of Norway being classified as
farmland, compared with 71 per cent of the UK, the overall

consumption of fertilizer is far lower, and any negative environmental impacts aren't as widely felt.

The Norwegian farming system involves rigorous monitoring of sheep performance and strict disease control measures. Lambs must reach a prescribed weight before they can be sold. Ewes that don't produce big lambs are dispensed with, so improving the productivity of the flock. Because of the high costs of keeping sheep inside, the aim is to sell off the year's crop of lambs in the same year they are born so that they carry as few as possible through the winter. Lambs that don't meet the required weight by the autumn might be given concentrated feed, but they generally get fat on the abundance offered by the hill. Sheep aren't closely shepherded, but are checked weekly, usually on a rota by a group of farmers who graze their flocks in a particular area.

Norway's total national flock numbers a little over two million sheep, a figure which satisfies Norway's national demand – imports and exports are a tiny part of the market. The UK has over 23 million sheep, many of which are exported. There are nearly as many sheep in Cumbria as there are in the whole of Norway. Although there are some significant differences, in many ways sheep farming in Norway is how farming used to be in the Lake District. It is a system designed to fit within the constraints of the landscape and the climate. Flocks are small, and so every animal gets more attention. Although fertilizers and concentrated feedstuffs are used, the main resource that the system relies upon is the natural growth on the hill. Keeping the numbers of animals low means that only a proportion of fodder from the hill is consumed each year, ensuring there is always a generous surplus for wild creatures.

There was scepticism among the Cumbrian farmers present at the Ambleside campus that hefting could function at the low sheep densities employed in Norway, despite the Norwegians' calm assurance that it absolutely could. During the question-and-answer session with the Norwegian farmers, I suggested that moving towards the sort of densities of grazing employed in their country might be a good way to restore the Lake District's upland habitats, while maintaining the cultural heritage and traditions of hill farming. Before I'd finished speaking, a hill farmer from the Langdale Valley, one of the most heavily grazed areas in the Lakes, stood up and asked, 'If you hate sheep so much, Lee, why are you even bothering with them?' He didn't really want an answer, and the discussion quickly moved on.

I've had this sort of 'sheep-hater' accusation hurled at me quite a few times. There was a time when comments like this would have been a huge blow to my confidence, but I'm used to them now, and I know where they come from. It wasn't that the hill farmer necessarily objected to the suggestion of reducing sheep numbers – more the fact that it was being made by me, someone who wasn't a 'proper' farmer. Nobody wants to hear an outsider suggesting that their way of life needs to change. But change is happening. I know loads of farmers who are reducing their sheep numbers for both economic and environmental reasons, and nobody's criticizing them. What I'm slowly learning is that trying to impose change on Lake District farmers is a pointless task. Change will only happen when farmers choose to bring it about themselves, and where the moral and financial support from the rest of society and from government makes it attractive to do so.

CHAPTER 10

The Alps

GRAN PARADISO:
Great Paradise
(Italian)

Norway's story is one of dramatic change, of nature given the time to bounce back from historic damage. Other European upland areas haven't had quite such a tumultuous time.

Spanning eight countries, the Alps are the longest, highest and botanically richest mountains in Europe, dwarfing our tiny hills in every respect. Despite the obvious physical contrasts, what the Alps and the Lake District have in common is that both are essentially agricultural landscapes. Traditional, light-touch farming is as distinctive a characteristic of the Alps as shepherds gathering sheep from the fells. Yet according to everything that I'd read, nature isn't faring anything like as badly in the Alps as it is at home.

With my sabbatical spent, I needed other ways to find out how this balance had been achieved. A family holiday to the Gran Paradiso National Park in the Italian Alps gave me the perfect opportunity to carry out more research, interspersed between pizza lunches, ice creams and cable-car rides. This was my first trip to the Alps. The snowy peaks, the tumbling rivers, and the bright blue glacial lakes are every bit as jaw-dropping as advertised. But it was the detail, the texture of the ground cover provided by the diversity of the plant life, that really blew my mind.

The holiday was a get-together for Becki's side of the family. Nature is a big thing for all of us: Becki's brother Chris is a seriously good birder, his wife Emily is a medicinal herb expert, my mother-in-law is brilliant at fungi and Uncle Tim and Aunt Ruth, who live below the mighty Ben Lawers, know a thing or two about plants and butterflies. My daughter Aphra is a budding botanist and Elliot loves moths and rocks. Becki and her dad are interested in all of it and how all these pieces fit together to make up the whole. But they get a bit frustrated at how very slowly the rest of us walk, stopping every two seconds to peer intently at things.

The highlight of the trip is a walk from Valnonty up to the Vittorio Sella Refuge. It's a serious hike, too much for the kids, who opt instead to explore the local waterfalls, and so, laden with optics and guidebooks, it is just Chris, Tim and I who hit the trail. With such richness of birds for Chris, butterflies for Tim and plants for me, it's something of a miracle that we make it beyond the car park.

Setting off from Valnonty, we start by walking through the communally managed St Ursus hay meadow on the valley bottom. This single huge field is looked after a bit like a Lake District common, with individual farmers each having rights to cut and graze small sections. Hay time is beginning, and little alpine tractors are arriving to take their allotted crop. Elsewhere in the meadow, small herds of cattle, replete with the obliga-tory bell, graze in tiny electric-fenced plots, tended by gangly, bored and smoking Italian youths. These meadows are recog-nized as a 'Wonder of Italy', in part for the continuation of their traditional, communal management. Like many old hay meadows in the UK, this one is on a floodplain and has a river running through it.

Floodplain meadows usually have deep, free-draining soils, fed with silt from the river when it floods. They are generally flat, making them easy places to cut and crop, and when man-aged in the traditional way are botanically one of the richest habitats in the temperate world.

The creation of hay meadows has been a gradual process. Prior to our taming of the wilderness, most floodplains would have supported a mosaic of tall herb fens, wet woodlands, swamps and marshes. Untamed rivers would have ripped through these habitats, changing course during times of high flow or to work around some natural blockage, a tree or a build-up of gravel. These valley bottoms would have been dynamic, with patches of habitat constantly being formed and destroyed, creating space for a vast range of plants and ani-mals. The first impact of people on these primordial floodplains would probably have been in the removal of trees, followed by the introduction of grazing animals, resulting in increasing openness. As people became more settled, beds of grasses, reeds, and tall wildflowers like meadowsweet, saw-wort, gypsywort, lesser meadow-rue and mint would have been cropped to provide winter fodder. Through steady incremental changes, farming made the meadows more uniform. Boulders were removed and marshes and swamps were drained to create larger, flatter areas that could be more easily cut.

The soils of traditional hay meadows are characterized by

having lower nutrients compared with those of modern agriculturally 'improved' grasslands, and perhaps counter-intuitively, this is part of what keeps them botanically rich. With an annual cycle of cutting and then grazing, nutrients are slowly stripped from the land into the bodies of the animals and the hay. Lower nutrients tend to favour flowers over grasses, and species like yellow rattle and eyebright, which parasitize grasses, give the flowers a further advantage. There is something of a positive feedback loop here; as the grasses are suppressed the sward becomes more open, creating more space and open ground for even more flowers to establish. The annual cutting, traditionally carried out after seeds have been set, helps to scatter the seeds across the meadow while at the same time giving them access to bare ground in which to germinate.

Hay meadows therefore start to resemble the conditions found in the mountains. Low-nutrient soils and reduced competition from vigorous species are two of the conditions to which mountain flowers are well adapted. The rivers provide a convenient means of transportation for seeds to travel from the mountains above down onto the floodplain, and over thousands of years this has helped many species to colonize these anthropogenic habitats. Our hay meadows in Swindale contain many species that are also typical of cliffs and crags. Wood crane's-bill, globeflower, smooth lady's-mantle, devil's-bit scabious, hay rattle and eyebright find themselves equally at home in the ungrazed, wild crags as in the tamed meadows below.

I can recognize many of Cumbria's flowers and a good proportion of the grasses, sedges and rushes, but then there are only 1,373 to choose from. Here in the Alps, there are over 4,500. Despite my best intentions and the help of a field guide to alpine flowers, I'm defeated by the diversity of the meadows spread out before me. Though I'm able to pin most of the flowers to a genus, it's beyond me to determine whether I'm looking at *Alchemilla fallax* or *Alchemilla fissa*. Time is short and the higher ground beckons.

⚜

For every 100 metres of altitude gained as you ascend a mountain, the temperature falls by about 1°C. The top of the Shard, Western Europe's tallest building, is 3°C chillier than down on the street below. The summit of Ben Nevis is on average almost 16°C cooler than the shore of Loch Linnhe at sea level, just four and a half miles away. As the temperature drops, the plants and habitats vary accordingly, each finding its own climatic comfort zone.

As we start to climb out of the beautiful but heavily modified valley bottom, the switchback path is soon surrounded by pine forest. These open woodlands are nothing like the serried ranks of conifers at home. The wide spacing between the trees allows light to reach the floor, and the glades are full of flowers and butterflies, while the local black variety of red squirrels scamper and nutcrackers jostle overhead. In proper mountain environments like the Alps, woodlands like these are highly valued, not only as a source of timber and firewood, but also as a defence against avalanches. A protective swathe of woodland almost always occupies the zone above towns and villages in the Alps, the trees acting as a natural brake to snow, which might otherwise annihilate a community.

As we climb, the pines shrink, and a zone of scrubby woodland takes over. Here, willows and juniper are hunkered down, well suited to spending large parts of the year buried in snow. Then, as the habitat starts to open up further, we pass a couple of ruined stone buildings that look like old shepherd's huts. When the slope levels off, the scrubby trees give way to open ground, but not like any open ground I've seen before. A carpet of colour stretches out before me. Every flower that I hoped I might one day see in the UK is weaved in an abundance that verges on extravagance. Great mounds of mountain avens, *Dryas octapetala* – a species that clings pathetically to two mountainsides in the Lake District – are interspersed with orchids, harebells, geraniums, gentians, eyebrights, wintergreens, fleabanes, saw-worts, dwarf willows, spurges, butterworts, succulent sedums, house leeks, clovers and vetches. Being conditioned to hills where flowers are often few and far between, this, my first taste of a real alpine meadow, is like some vision of naturalist heaven.

Everywhere, amid the din of a million grasshoppers, are clouds of butterflies in every colour. Like the flowers, there are just too many species to identify, so I make do by naming them as fritillaries, blues, mountain ringlets, metalmarks or nymphalids. At the edge of the path, we come across the comically lurid and gigantic orange and black caterpillar of a spurge hawkmoth. It looks more like a tropical species than something you'd find on a rocky European mountaintop.

Mountain Avens

Butterflies and moths broadly speaking need two things from the kingdom of plants: a food-plant for their caterpillars to eat and flowers providing nectar for the adults. Many species have a specific relationship with their food-plant. The adults will lay their eggs onto it, so that as soon as the first instar caterpillars emerge they can get on with the important job of eating it. The spurge hawkmoth caterpillar is on a spurge plant, the brown species are probably seeking out a grass to lay their eggs on, some of the fritillaries will be after violets. Most caterpillars will only eat the species of plant their mother laid their egg on, genetically predisposed to it through aeons of co-evolution. If you swap the egg of a large white butterfly from the cabbages in your garden with the egg of a small pearl-bordered fritillary from the leaf of dog violet, neither caterpillar will survive long after hatching.

After they've gone through various stages of larval development, cocooned themselves and emerged as an adult, butterflies become slightly less choosy. Any flower that a butterfly can uncoil its long, spring-like proboscis into to lap up some nectar will do. Members of the daisy, thistle and scabious families are favourites for many species. In return the plant gets a pollination service, with the butterflies being duped into ferrying pollen from flower to flower. So, diverse and abundant flowers result in diverse and abundant butterflies, which help pollinate the next generation of flowers. These butterflies, along with all the other, less glamorous insects, form half of the next link in the food chain.

The most visible birds in these alpine meadows are the choughs. We see red-billed choughs, which maintain a tenuous foothold in the UK, as well as yellow-billed alpine choughs, both of which are specialist invertebrate feeders. As well as caterpillars, they feed on beetles, grasshoppers, snails and other insects that they glean from the floral carpet. Choughs are big birds, about the size of magpies, requiring a lot of food, which this landscape clearly delivers in spades.

❧

Cresting the lip of a hanging valley, we see the *rifugio*, a mountain cabin, up ahead. I hadn't really given much thought to what it would look like. I suppose I was half expecting something small and ramshackle, not this huge, serious-looking structure equipped with heat, light and power. The effort in building something of this scale up here, where presumably the only option was to fly in the materials, amazed me. What I'm most pleased about is the fact that there is a bar with a roaring fire, proper Italian coffee, hot chocolate and beer. We'd set out in beautiful, warm summer weather down in Valnonty. We've gained 1,000 metres of altitude to reach the *rifugio* at 2,588 metres, and in doing so we lost 10°C of warmth, a situation I'm acutely aware of in my shorts and T-shirt. After some time warming up near the fire, followed by half an hour cooling off again while watching marmots and some distant ibex, we head back the way we came.

Reaching the lower elevations of the flowery plateau a couple of miles from the *rifugio*, we pass a small but comfortable-looking seasonal dwelling, which must be the modern replacement for the ruined stone building we'd seen on our way up. Nearby, the current resident is herding a small flock of dusty sheep whose bells are clanging cheerily. My first sensation on seeing them is mild horror – how dare these sheep even consider eating all these fabulous wildflowers? Then I look around. There are about thirty sheep in this little flock, and they are the only ones we've seen during a whole day of walking. Clearly this is a very different type of hill farming to the sort I'm used to see-ing at home. These sheep are only here for the short season between snowmelt and snowfall. At this low density, their dis-turbance makes a positive contribution to maintaining the rich conditions, preventing any one species from becoming too common, and creating gaps for new plants to grow. Being actively shepherded, the sheep are better cared for, less likely to wander off a cliff, and the shepherd will be able to spot any sickly members of the flock much sooner than animals that are left to fend for themselves for weeks at a stretch.

Somehow Chris spies a lammergeier on the opposite side of the valley as we descend through the pine forest. God knows how he spotted it. He's one of those committed birders who seem to have a near supernatural ability to see things that the rest of us miss. We watch it through his telescope for a while, not totally sure it isn't a discoloured rock until eventually it proves Chris right by soaring off into the void. Lammergeiers, or bearded vultures, are one of the few species on the planet whose diet is made up almost exclusively of the skeletons of dead animals. This is a landscape so rich and intact that even its bones support fascinating creatures.

❧

I'm under no illusions that the Lake District ever was, or ever could be, a lot like the Italian Alps. The far greater height of the mountains and their location on the globe means the cli-mate is very different, with warmer summers and much colder, snowier winters. The geology is also distinct, favouring a dif-ferent suite of species. However, there are lessons here for those who want to learn. Like the Lake District, this area of Europe is a cultural landscape. Farming has played a part in shaping the scenery and continues to be a vital component of the rural economy with locally distinctive farm produce, cheese, meat, honey and fabrics selling at a premium, partly into the thriving tourist industry. In both regions, tourism is now economically far more important than farming.

In 2018, 19.38 million tourists visited the Lake District, spending £1.48 billion and supporting over 18,000 full-time equivalent jobs. These are astonishing figures, when you consider that only around 42,000 people live in the area, a very large proportion of whom are retired. Getting hold of facts and figures about farming is slightly harder, but according to the State of the Park report in 2018, there are around 2,600 people employed in farming across the national park on about 1,200 farm holdings. Farming and tourism are heavily intertwined, with many farms relying on tourism for part of their income, and many tourists enjoying the visual and cultural appeal of the farming.

Like all mountainous areas, the Alps and the Lake District have always been and always will be agriculturally marginal, with landforms and climatic conditions rendering them unsuited to intensive modern agriculture. Prior to the Second World War, upland farming across Europe was probably broadly consistent, being just one part of a more diversified rural economy. After the war, the economic paths of mountain regions within Europe and the UK diverged. In the UK, the uplands were just a hilly region playing their part in the national effort to boost production, with the introduction of subsidies pushing numbers of livestock up to a level that took a major toll on the natural environment. Some mountain areas in Europe, on the other hand, focused instead on tourism, particularly skiing. With an economy more based on recreation, and with less of a drive towards agricultural intensification, habitats were less damaged, though ski runs create their own environmental issues.

The Alps is a large region spanning multiple national borders, so there is no single approach to financial support for farming. In Switzerland, the nation dominated by the Alps more than any other, farming subsidies are very generous and tend not to be focused on production. The average Swiss farm receives €46,000 per year, almost three times as much as the average farm in the UK. These payments reward Swiss farmers not for producing food, but for delivering public goods, for keeping rural traditions alive and maintaining a landscape that is both beautiful and healthy.

When we proposed removing our sheep from an area of common land around Haweswater's southern tip to give the plants an opportunity to spread out from their refuges on Harter Fell, we caused a ruckus. A series of fraught meetings with the Federation of Cumbria Commoners ensued. We were told in no uncertain terms that our action would undo thousands of years of custom and practice and could risk destabilizing the delicate balance of hefting across the whole of the Lake District. One incredulous farmer said, 'You can't take sheep off the hill just to protect a bunch of bloody flowers!' I guess he believes that the land is for livestock and that to do anything other than use it for producing food would be wasteful. I believe that managing a landscape for flowers, and the mass of life that relies on them, isn't a bad idea either. I challenge anyone who's spent time in an alpine meadow and experienced the rush of joy that it gave to disagree. These two world views don't have to be mutually exclusive.

CHAPTER 11

Wild Lakeland

RIVER LIZA:
The bright, light river
(Old Norse)

Trips to Norway, Scotland and the Alps have been revelatory, but I've learned a lot closer to home in recent years too. There's a genuine movement for change building in the Lakes. The membership of the group of conservationists that I meet up with every few months is constantly growing, and our get-togethers are always fizzing with energy and ideas. The number of nature-friendly farmers is undoubtedly on the rise too. Whether it's because they can see that farming with nature is likely to make good financial sense thanks to the new direction of government incentives, or because of an implicit desire to help the environment, or perhaps a bit of both, there are increasing numbers of people doing positive things for wildlife on their land. Sure, there are still those who deny the need for change, but my sense is that they are a dwindling minority.

There will always be extreme views within both camps, but I'm seeing the progressive middles of the two 'sides' of the debate about land slowly and surely coming together. In her book *Animal, Mineral, Radical*, B. K. Loren defines community as having 'little to do with like minds. It has to do with very differently minded people finding a way to get along because we all live in, are connected to, and share a sense of place.' By that definition, if the trend continues, there will be no farming community or conservation community in a few years' time – there'll just be a community.

Just as there is a spectrum of opinions among the farmers and land managers who live and work in the Lake District, there is also a spectrum of projects and approaches. Our group's main purpose is to visit sites that will teach us about land that is working for nature and for people. We've visited many of the brightest spots together over recent years, trips that have been as insightful as they've been inspiring.

At the wildest end of the spectrum is Wild Ennerdale, England's longest-running rewilding project. Sitting at the opposite corner of the Lakes to Haweswater, in the far western fells, Wild Ennerdale is run by a partnership of four organizations, the National Trust, Forestry England, United Utilities – owning 4,400 hectares of land in the valley – and Natural England.

My first visit to Ennerdale was a few months after I started at Haweswater, part of an induction to United Utilities' projects that my friend John was giving me. Ennerdale Water, sitting club-shaped and blue at the bottom of the valley, is United Utilities' main interest. During the last century it supplied much of West Cumbria's drinking water. Although it's a natural lake, a modest impoundment was constructed in 1902 to maintain its level and keep water reliably flowing into an intake pipe, which ran to the treatment works, and then to people's houses.

Ennerdale has always been United Utilities' cheapest single supply, requiring the least treatment to get the water to the standard needed for human consumption. Standing on the shore, where John and I started our walk up the valley, the depths were far clearer than in many of the Lake District's other, murkier water bodies. Walking along the lakeshore, there was a palpable sense of purity that set the place apart, as if the water's clarity were somehow transmissible to the air.

Ennerdale's crowning glory is the River Liza, which flows into Ennerdale Water's eastern end, over a broad gravelly delta. The Liza's name comes from the Norse, and means 'the bright, light river' – it lives up to it. All along its five-mile length, smooth cobbles glow beneath crystalline water. The pale pink of the Ennerdale Granophyre and the grey of Borrowdale Volcanics, the valley's two main rock types, give the river its distinctive palette. Flowing from the high fells that encircle the valley, the Liza is a river of fearsome power. Its rushing torrent keeps the riverbed in constant motion, making it hard for plants and algae to colonize, and so the gravels stay clean and clear.

Perhaps because the valley was so sparsely populated, or because the river was so powerful and unruly, the Liza was spared the straightening, dredging and banking that have tamed so many other rivers in the UK. The approach of the Ennerdale partnership is to allow the river to carry on doing exactly as it pleases, and the river alone is enough to earn the valley its Wild moniker. It braids and wanders, erodes and deposits, ever-changing in its form and what it offers to wildlife.

Adapted to deep, cold water, Arctic char, a relative of salmon and trout, spawn in the Liza's extensive gravel beds. They are one of Britain's rarest species of fish and, until a few years ago, Ennerdale's population was on the brink of extinction, their numbers down to just a few hundred. Removing barriers that blocked upstream migration, together with a captive breeding and release programme, has helped Ennerdale's Arctic char to thrive again.

The river is constantly reinventing itself. Ephemeral deposits of gravel are colonized by seeds washing down from upstream, forming seasonal flowering cushions. Many of these get smashed and reconstituted further downstream during the winter storms. If they stick around long enough for larger, woody species to establish, roots give them stability, and they can grow into small scrubby islands. The interplay of land, water and plant means that all stages of this succession are represented, in a state of constant dynamism.

The Liza is a wonder, a vision of what many of our upland rivers used to be before we started meddling with them. The valley it flows through is not so pristine. In response to timber shortages resulting from the First World War, the Forestry Commission, established in 1919, planted tens of thousands of hectares of commercial woodlands across the UK. Many of these were in upland areas, where marginal hill-farming interests were swept aside to make way for the greater good. Ennerdale came under the Commission's gaze in the late 1920s, leading to 1,200 hectares of new plantation, comprised mostly of closely packed non-native conifers, smothering the valley's floor and lower slopes.

In the intervening decades, while the conifers slowly matured, the technology of timber harvesting modernized. To be commercially viable, plantations must now be felled and extracted quickly and efficiently, which today means using large forest harvesters. These mechanical leviathans can fell an adult tree, remove its branches and chop it into prescribed lengths in under a minute. A forestry tractor, or forwarder, then conveys the sections of timber to the roadside, ready for delivery to the sawmill.

For all this to work smoothly, there needs to be good access, both to get the monstrous machines into the forest and to allow the trucks to take the wood away. The lanes that wind through sleepy villages to Ennerdale couldn't really be less suitable, so the Commission found itself with a forest that was largely unharvestable. It was this, and a measure of luck that the right people were in the right jobs, that prompted the partners at Ennerdale to start thinking differently about the valley's future.

Forestry still dominates much of the view at Ennerdale, especially from the valley bottom. Although some tree felling is still happening, much of the woodland is being left to age and rot and, where it's within striking distance, to fall into the Liza. The interaction between big chunks of fallen wood and the river is a fascinating one. In most places, our compulsion for tidiness means that trees are quickly whisked out if they fall into water. Not so in Ennerdale. The Liza now has lots to chew on. Woody blockages force the water to forge new channels across the floodplain. Although you could say that these particular trees aren't natural, having been planted here as a commercial crop, the process that they stimulate most certainly is, and the river is more alive as a result.

Unlike more typical nature conservation projects, the Wild Ennerdale partnership isn't concerned with creating or restoring habitats directly. The focus of their gentle stewardship is kick-starting the underlying natural processes that will lead to self-willed recovery, giving nature the freedom to do the restoration work herself.

As John and I made our way up the Liza, splashing through shallow water around thickets of gorse growing in the lee of fallen trees, another natural process was about to introduce itself. We'd sat for a bit of lunch on a fallen spruce log, half in the water, half stranded on a gravel bar. Chatting about the river, lamenting that it was the only one of its kind in the Lakes, we noticed the sound of something big coming our way through the undergrowth. The crashing and snapping of branches grew louder, gorse and willow being shaken by whatever it was approaching. Surrounded by conifers, a broad braiding river at our feet, there was something distinctly Canadian about the scene, and I began to wish I'd packed my bear spray.

The beast that emerged from the scrub wasn't a bear, though it was nearly as hairy. A thickset, woolly-eared black Galloway cow crashed into the gravelly clearing. It stooped for a drink from the Liza, waded across and without a glance in our direction, vanished into the undergrowth on the other side. It may only have been a cow, but there was something thrillingly wild about the encounter.

The cow John and I saw was one of a small herd of Galloways that roam through Ennerdale's tracts of forest and fell. Their owner, Richard Maxwell, is another vital player in the Ennerdale team.

I met Richard a few months after his cow had introduced herself, at an event to celebrate Wild Ennerdale's ten-year anniversary. Richard was telling the group who'd come for the day the story of his cattle and the impact that they'd had on nature and on him. Gentle and affable, Richard is a Cumbrian farmer born and bred. Dark-haired and broad-shouldered, a bit like his cattle, Richard's connection to Ennerdale is deep and strong, built up over a lifetime of work in and around the valley. As a younger man, his main interest was in commercial cross-bred sheep and continental cattle, which he had farmed intensively, as was the norm for the time.

Over the years, the Wild Ennerdale partners had grown increasingly convinced that the grazing of free-roaming, large herbivores was a natural process that was missing from the valley. In the early 2000s they approached Richard and the other local farmers to see if they might be interested in putting cattle into a 140-hectare block of recently felled forest. This was a pretty outlandish concept; forests aren't usually grazed in the UK, especially not conifer plantations. Perhaps the offer of a slice of a government stewardship grant helped to persuade him at first, and Richard decided to give it a shot, investing in a small number of hardy black Galloways for the purpose. A couple of years later a further 250 hectares were added, then 500 more shortly after. Today, the cattle wander freely over 1,000 hectares of what must be some of the most unlikely-looking livestock pasture in the country.

Richard has never looked back. As he stood with his curly-headed cattle, telling us about the challenges of tending to them in such a massive and wooded landscape, it was obvious that he isn't just in it for the money. He clearly loves his cows, and not only because they earn him a living through meat sales and subsidies, but because they're characters that he's helped bring into the world and spent time with.

Embracing change in the way that Richard has done takes real courage, and he had to endure ridicule from his farming peers at first, all convinced that the cattle would die and that such an extensive system couldn't possibly work. Richard's Galloways grow more slowly than cattle in more commercial setups, but meat is only one of their functions now, so this is a workable compromise. These cattle stand in for the missing wild herbivores that the rest of the valley's wildlife evolved alongside.

As Richard's cattle wander around the valley, crashing through patches of scrub, woodland, bog and meadow, their shaggy coats pick up seeds. Plants like water avens have evolved a mechanism for seed dispersal that relies entirely on animals like cattle. By late summer, water avens' nodding flowers have developed into bur-like balls of seeds, each armed with a sticky, hook-tipped barb. As animals pass by, the tiny hooks grip to fur as efficiently as Velcro – a product inspired by sticky seeds very like those of water avens – and the hitch-hiking seeds are carried away. As the hairy beast scratches or wades, shakes or moults, the seeds are dislodged, likely a long way from where they started out.

Other plants catch a ride inside the cattle. Seeds protected with tough coats, like the stones of cherries and members of the pea family, like gorse and broom, can pass through a cow's gut and come out unscathed at the other end, conveniently deposited with the cowpat, a ready-made, nutrient-rich bed. The cowslip, one of our cheeriest spring flowers, is named because our forebears noticed that it regularly grew where cowpats had been, in the places where the cows had 'slupped'. Aside from transporting seeds, heavy hoofprints and the disturbed ground where cattle congregate expose the soil, creating gaps in which seeds can grow into new plants, preventing any one species from taking over.

It's a slow process, but the cattle in Ennerdale are starting to blur the boundaries between forest, field and fell. Where the conifers have been chopped, broadleaf trees are appearing, and they're also spreading up the valley sides, diversifying the heather and grass-clad slopes. The whole valley feels flowery and rich, and we have the cattle to thank for their part in it.

Richard still keeps sheep, but he has a lot less than he used to. The conversion to extensive, nature-focused farming that now makes up the bulk of his day-to-day working life is much more manageable than when he had hundreds of sheep to keep alive. Richard told the group attending Wild Ennerdale's ten-year anniversary that the switch had allowed him to get to know his family for the first time. He even takes holidays, something he never did before. Richard is living proof that change can be good, on all sorts of levels. Two of his neighbours, both of whom thought he was mad at first, now also help to graze cattle in the valley.

Wild Ennerdale is the largest area in the Lakes where natural processes are given priority, but there are other pockets of wildness that can be found if you take the time to look. A little way east of Ennerdale, towering over Bassenthwaite Lake just north of the bustling tourist town of Keswick, is another Forestry Commission woodland called Whinlatter. More accessible than Ennerdale, Whinlatter is still a viable commercial site, producing timber, as well as offering adventure to visitors. Hidden by the forest, far from the trails, play area, café and Go Ape aerial assault course, is a remnant of extraordinary habitat.

Not all the land that the Forestry Commission acquired was suitable for planting trees. Wet areas were often drained, but there was little that could be done to encourage the rocky hilltops or the steepest slopes to produce a harvestable timber crop. Above the highest of Whinlatter's conifers are several islands of open ground that have been left to their own devices for close to a century.

Jean, who works for Natural England, suggested that our group should visit, so ten of us trooped up one July day, with Gareth from the Forestry Commission leading the way.

Following one of the forestry roads up from the visitor centre, with the occasional mountain bike whizzing past us, everything was neat and ordered. The well-made track between rows of towering conifers made for a pleasant shady stroll, as we traded stories of what we'd all been up to for the past few months. The trackside verges were full of flowers, and the shelter of the trees made life easy for the insects feeding on them. Further in, things were less rosy, and little plant or insect life survived in the deep shade of the plantation proper.

After working up a bit of a sweat, we turned off the forest road and plunged into the darkness under the conifers, pushing our way past scratchy lower branches like the children passing through the wardrobe into Narnia. Sure enough, the world on the other side was completely new. The remains of a long disused path ran along a ridge in front of us. To the right was the dark wall of trees, mussed by the wind; without the protection of the forest, many had fallen into the open space of the hillside. To the left, the ground fell away steeply, too sheer for planting. Other than a scattering of self-seeded larch and spruce, it was thick with purpling heather and blotched with vibrant green bilberry. Our progress slowed as we stained our fingers and sweetened our tongues with the ripening berries.

We rounded a small rocky outcrop and walked into a habitat that was more soft play centre than mountain. On north-facing slopes in hyper-oceanic climates like the Lake District, a unique and slightly bizarre habitat known as Atlantic heath can develop. The high rainfall and the cool shady aspect give mosses the upper hand over shrubs, flowers and trees, and they grow into gigantic springy hummocks. I sank my arm into one, burrowing my fingers down into the cool, humid moss right up to my shoulder, and still didn't reach anything I could identify as the ground. So huge and spongy are the hummocks that it isn't really possible to walk through Atlantic heath, instead you have to employ an undignified combination of leaping, crawling and falling. Gareth had to leave early for a meeting back at the visitor centre, and rolled and bounced off down the hill, leaving us sinking into mossy nests with our lunches, laughing at how Tigger-like he was.

The bulk of those hummocks were made of sphagnum, a moss famous for its incredible water-holding abilities. This surreal hillside acts as an enormous sponge when it rains, massively slowing the flow of water and reducing the risk of flooding for those living below. Even though there's not much feed value in moss, the increase in livestock numbers has almost certainly reduced the extent of Atlantic heath in the Lakes, as even modest sustained trampling and dunging can damage it. Historically, Atlantic heath would have been much more widespread at the tops of steep, damp hillsides, where grazing animals would have previously been fewer and farther between. Only in areas like the accidental refuge above Whinlatter has it been able to hang on.

We had a couple of hours to spare in the afternoon, so while everyone else headed back to their offices, Pete, who works for the Woodland Trust, marched me up into Coledale, a long valley just below Whinlatter. There, Coledale Beck wiggles its way along the valley bottom, which on that July day was a gently tumbling stream, burbling unassumingly over its rocky bed. Below the towering crags at the valley head we could see the remains of Force Crag Mine, which had extracted lead and other minerals up until 1991.

Braithwaite, the village at the bottom of the valley, has

suffered serious flooding on numerous occasions in recent years. Heavy rain falling on the steep slopes with their short covering of vegetation triggered landslides, and so the torrents that swept into the village took massive quantities of soil and rocks with them. As part of a package of responses designed to reduce flooding in Braithwaite, the National Trust, which owns Coledale, worked with Pete, Natural England and the local farmers and residents to plant trees in the valley. I'd visited on a foul day a couple of years previously, soon after the trees had gone in the ground, and I was keen to see how they were doing on a day when I could stand upright.

Pete's been tremendously successful in recent years, in part thanks to the flooding, which pushed tree planting up the agenda. All of a sudden, trees were not just about wildlife or carbon capture, they were part of wider efforts to keep people's insurance premiums down and perhaps reduce the government's need to pour concrete in costly engineered flood-defence schemes.

The Woodland Trust doesn't have much land of its own in Cumbria, so Pete's job is all about trying to make it as easy and attractive as possible for farmers and other landowners to get trees in the ground. He's a bit of a dynamo. An incomer like me, Pete seems to know just about everyone, and has an uncanny ability to sniff out opportunities for tree planting. Over the past decade he's oiled the wheels of projects that have resulted in the planting of over two million trees across Cumbria, which in addition to environmental benefit often also provide an important source of grant income for farmers.

There's lots of evidence to suggest that landscapes with trees help to reduce flood risk, compared with open terrain. Trees and the thicker vegetation that develops around them if livestock are excluded increase the ground's roughness, making it harder for water to flow over the surface. Just as the pine forests in the Alps protect mountain villages from avalanche, trees can also act as a brake to earth and rock released in a landslide. Their roots help water to get into the ground more quickly and to penetrate to greater depths, as well as physically binding the soil in place.

Trees are close to a silver bullet. That they can reduce flood risk, and lock up carbon, and support more wildlife seems almost too good to be true. Establishing them at the scale needed to make a material difference is no easy task, though, as Pete was about to show me.

The trees had been in the ground for four years, but it was hard to spot them from any distance away. It wasn't until we dropped off the track that used to convey the lead, zinc and barites from the mine down the valley that they started to appear. Pete pointed at the places where some had been lost in a more recent landslide, and where earlier planting had failed because the trees hadn't been given adequate protection from grazing. We climbed over a fence into the hard-won plot where most of the planting had taken place. Not everyone wanted trees in the valley, despite the well-documented benefits that they'd bring. Some local people were concerned about the loss of grazing land to the woodland, others worried about the visual change, that the trees would alter the valley's character. After much wrangling, an area on the valley bottom was agreed upon, fences went up to keep the sheep out, and the first trees were planted.

Pete pulled out a folding saw and started lopping off the tops of some of the leggier trees. This seemed a bit brutal, but he explained that the miniature greenhouse effect caused by the tree guards had made some trees grow too fast, leaving them vulnerable to being snapped in strong winds. Pruning the tops of these upstarts gave them more strength lower down, so that when the guards were removed in a few years, they'd have a better chance of staying upright.

Pruning as we went, righting trees where their wooden supports had rotted, removing guards from trees that had died, Pete could assess how successful the planting had been. We came across several happy-looking clumps of aspen, which was encouraging. A relative of willow, aspen is highly palatable to grazing animals. It has lovely crenelated leaves and a flattened leaf stem, which cause the leaves to flutter in even the gentlest breeze. According to folklore, this trembling is the tree's way of showing its remorse for its timber having been used to make the cross that Christ was nailed to. On one healthy specimen we came across a trio of extraordinary puss moth caterpillars, one of many insects with a liking for aspen. Vivid green and black with a sinister red-ringed clown face at one end, and twin tails with extendable red filaments at the other, puss moth caterpillars have evolved to look as frightening as possible. If their appearance isn't enough to scare off would-be predators, they can also squirt formic acid from their thorax, the same toxin that gives ant venom its bite.

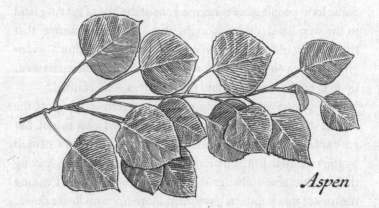

Aspen

Aspen is a species that suckers, new stems growing up from lateral roots just below the soil. In the American state of Utah a stand of aspen has managed to grow into truly gigantic proportions. To the casual observer, the forty-three hectares of pale grey aspen trunks are just a patch of forest like any other, but in the fall all their leaves begin to turn at exactly the same moment, making the whole hillside yellow. It turns out that these 40,000 trunks are all connected: they are suckers growing out of the same root mass. The whole stand is a single organism, the largest, heaviest and oldest plant on the planet. It has been named Pando, the Latin for 'I spread out', but it's also locally referred to as the trembling giant. Above ground, aspens live fast and die young – Pando's individual stems only persist for around 150 years each, but its root system is prehistoric. Nobody is quite certain, but current estimates put it at around 80,000 years old. Pando is a relic from a different epoch, and Utah's climate today doesn't suit it. It's thought that Pando hasn't flowered or produced seed for thousands of years and its survival now relies entirely on its ancient roots sending up new stems to replace those that die.

Tragically, after tens of thousands of years, Pando is in trouble. The saplings that rise from the roots are extremely palatable, and an increase in grazing by wild deer and domestic cattle under Pando's dappled shade have removed generations of trees that would otherwise have stood ready to replace the veterans once they succumbed to old age. The whole gigantic stand is now comprised only of old trees. Unless something changes, there's a risk that when these die there will be no above-ground growth to send sugars down to support the root system, and that this living monument could wither.

The aspens that Pete and the local volunteers had planted weren't quite approaching Pando proportions, but they were starting to put up new suckers. The originally planted trees with their protective guards were safe, but new trees were rising unprotected. We reused guards from other trees in the area that had died, putting them over the new shoots to create curious-looking clusters.

The fences around the new woodland should have rendered the guards unnecessary. Unfortunately, keeping sheep on the right side of even the most well-built fence is a challenge. As Pete and I walked around the exclosure, we spotted a Swaledale ewe and her well-grown lamb that had forced their way under the water gate at the upstream end. With a bit of arm waving and shouting, we managed to chase them back out the way they came, and set to bodging a repair with cable ties and boulders that would have to suffice until Pete could return to do a better job. These break-ins are a common occurrence. Deep in their genes, domestic sheep are still mouflon, one of the nimblest of mountain creatures. Hill breeds like Swaledales and Herdwicks are more than capable of jumping four-foot fences when they feel like it, especially when the grazing on the other side is worth the effort.

Pete, the local volunteers, National Trust staff and others have returned many times over the years, adding new trees and replacing those that have died. Pete reckons there are about 3,000 well-established trees now, and the area is beginning to show signs of further natural recovery. Despite repeated incursions by sheep, the plant life in the exclosure is responding positively. Bog asphodel and cranberry are appearing in the boggy areas, along with lawns of sphagnum moss. Yarrow, bird's-foot trefoil, thyme and sneezewort pick out the drier areas, and willow and birch seedlings are popping up in the bare ground left by the landslides.

We need more trees in the fells, but getting them growing is a long, slow process. Even when agreement to plant them has been secured from the landowners, graziers, residents, local authorities and Forestry Commission, getting them in the ground is only the start of the job. Years of aftercare are often required for the trees to survive. The fences need to be maintained, the guards kept in place and then removed when the trees have outgrown them. There will be a lifetime of painstaking and demanding effort for my colleagues, our volunteers and me to ensure the survival of the tens of thousands of trees that we've planted at Haweswater.

Natural regeneration, without the need for planting, guards or fences, would not only be cheaper, but would also result in more natural, healthier woodland in the longer term. Trees are much better at choosing the right places to grow than we are at choosing for them, but natural regeneration is only an option in a limited range of situations. There needs to be a local seed source as well as a suitably low level of grazing, and the right soil conditions. In the fells, these conditions are rarely met, and so planting, with all the associated paraphernalia, is often the only practical option.

An even more essential consideration is that somebody needs to take responsibility for ensuring that the trees become established. Planting trees purely to access grant funding isn't likely to produce a good result. Farmers, landowners, residents and whoever else has an interest must really want the trees if they are to stand a chance of turning into a genuinely valuable natural resource.

❧

The Coledale Valley is part of the 20 per cent of the Lake District National Park owned by the National Trust. Most of their land has farming tenants, and much of it is common land, so their approach has to be inclusive, and is often slower as a result. Nevertheless, in the places where their tenants share their aspirations for nature, the land is becoming richer and wilder.

Our next-door neighbours at Haweswater, the sprawling Lowther Estate with its fairy-tale castle and polar explorer owner, Jim Lowther, are also taking major strides towards a more nature-rich future. They have reintroduced beavers to their land, reconnected floodplains, restored meanders to straightened rivers, planted thousands of trees, and replaced a huge flock of sheep with a scattering of hardy cattle and pigs. Their ambition is as enormous as their landholding, and to know that their work and ours will together add up to an even larger, connected area where nature can recover makes it all the more exhilarating.

Cumbria Wildlife Trust are doing great work too. Eycott Hill, a couple of hundred metres down the road from my house, is a fantastic nature reserve, growing richer each year. They have actively restored huge swathes of flower-rich hay meadows, planted trees, scrub and hedges, restored areas of bog by blocking artificial drains, dug ponds, and improved access, signs and interpretation for visitors. Their herd of hardy cattle play a similar role to Richard's in Ennerdale.

These institutional approaches to nature conservation, Haweswater included, are all a valuable part of the overall picture and collectively add up to around 5 per cent of the national park's area. The largest proportion, however, is in the care of the farmers. Whether through ownership or tenancy, it is agriculture that decides the future of the Lake District's wildlife.

CHAPTER 12

Farmed Lakeland

MATTERDALE:
The valley where the bedstraw grows
(Old Norse)

The average farmer doesn't have a communications team to send out newsletters or produce an annual report, and so the environmental work carried out on farms is often less advertised than that implemented by institutions. As I got to know more farmers personally, and made the effort to visit their farms, I started to get a better idea of what made some of them tick, what their fears and motivations were, and how much unsung good work was being done.

I'd been stupidly slow realizing how important such a simple act as meeting people was. Being an incomer, and not naturally very sociable, it was a struggle to pick up the phone, or to call round and meet the neighbours, which is what I should have done on day one. It's still not something I find very easy, but I'm doing it anyway.

The first local farmer that I got to know well was James Rebanks, Herdwick shepherd and best-selling author. James had got in touch to discover more about what we were doing at Haweswater while researching his second book, *English Pastoral*. His valley, Matterdale, is a higgledy-piggledy place that extends south from the Penrith to Keswick road to the shore of Ullswater, just a few miles from my home. We talk a lot and although we disagree about some things, we agree on more.

The transformation that James has made on his farm is remarkable. Miles of new hedgerows, huge new ponds and wetlands, rewiggled (his term) becks, restored hay meadows, a switch from just sheep to cattle, pigs, chickens and sheep, and ditching the fertilizer – they all add up to an exemplary nature-friendly farm. As his farm habitats are becoming more diverse, wildlife is returning. Little egrets and green sandpipers appeared soon after his most recent batch of ponds were finished. Strips of longer ungrazed vegetation along the streams and hedges that break up his fields are heaven for voles and other small mammals, and the tawny and barn owls that feed on them are increasing in numbers as a result.

James is as passionate about farming Herdwick sheep as he is about doing what he can for the environment on his farm and encouraging others to do the same. Furtively mentioning his growing fascination with birds, flowers, soil and trees to his neighbours, he said, was a bit like coming out. Finding out how many of them felt the same way was as much a surprise as it was a pleasure.

James has teamed up with his friend Danny, a mechanic whose family-owned garage is halfway between my house and James's farm. Danny has always been fascinated with nature, and with rivers in particular. Over the years he's made a whole range of improvements to the land around the garage. He's transformed stretches of the beck, planting trees and putting in small dams to slow the flow, like a human beaver. He's encouraged his neighbours to graze in a different way, giving land periods of rest, benefiting the flowers and insects.

In 2015, Danny's home village of Glenridding, on the shore of Ullswater, was one of the most prominent casualties of Storm Desmond, a record-breaking extra-tropical cyclone that caused misery across the UK. The flooding was devastating – it seemed as if half the mountainside above had been washed into the village, burying the road and cutting off phones, water and electricity for days. It took diggers weeks to shift the tens of thousands of tonnes of gravel and rock that had been brought down, and it was years before the village was fully functional again.

In response, Danny set up a Community Interest Company, giving him the ability to raise funds to provide natural flood relief through tree planting, 'leaky dams' in streams and better care of soils. Teaming up with James and a growing number of other farmers in the area, their collaborative efforts are really starting to make a difference. Because it is farmer and community led, they are having a huge impact – one it would be very difficult for the RSPB, with all our institutional baggage, to emulate.

Danny, James and their neighbours work with a whole range of organizations. Pete from the Woodland Trust is in the mix, of course, having spotted an opportunity for tree planting. Eden Rivers Trust have helped out with river restoration; Rob, the botanist who joined me for the survey of Rowantreethwaite Gorge, has contributed his considerable knowledge to help with plant surveys and meadow restoration; Natural England have provided funding and advice. It's a proper team effort, but the most inspiring thing is that this is led by the

farmers themselves. Without the bureaucracy of a large organization to wade through, James and Danny can get on with
making changes as they see fit.

⤺

Just over the hill from James's place, Sam and Candida Hodgson have been ploughing their lives into Glencoyne Farm,
where they've been tenants of the National Trust since 1995.
Set back from the road that skirts the Ullswater shore, which
in the summer can be nose-to-tail with tourists, their seventeenth-
century farmhouse is shaded for much of the year by the peaks
that form its spectacular backdrop. It's one of the largest hill
farms in the Lake District, extending to 1,258 hectares, roughly
the same size as the London Borough of Islington but, during
the winter months at least, with far fewer people.

Glencoyne is a bit like Naddle Farm in that it has a tiny
amount of productive land to support a massive expanse of
much wilder, woodier, wetter ground. Sam and Can are
doing brilliant things with their farm, stretching the definition of what modern farmland can look like. Can showed a
group of us around their wood pasture a few summers back.
Their cattle have free rein to wander inside a huge block of
land, extending from the lakeshore road up onto the lower
slopes of the fells above. There are dense clumps of trees here
and there, but mostly the land is patchy and chaotic, with
hoary alder, hawthorn, crab apple, oak, ash and holly scattered within a matrix of meadowsweet and bracken, grass
and rush, heather and bilberry. 'Glencoyne' means either
reedy or beautiful valley. It's not that reedy any more, but it's
certainly beautiful.

When managed well, wood pasture like this is perhaps the

ultimate union of farming and wildlife. The trees, which are allowed to reach a venerable age, would traditionally have had their branches periodically lopped, to provide 'tree hay' for winter fodder or for building or burning. This regular cropping, known as pollarding, doesn't harm the tree and new branches quickly regenerate. The main stems of these pollards provide holes for pied flycatchers and redstarts, and they support a diversity of lichens that can only grow on large open-grown trees. The low-intensity grazing in a wood pasture ensures that enough trees can grow to replace the veterans as they succumb to old age and allows a great bulk and variety of plants and insects to thrive.

Given the choice, all livestock like a varied diet. Cattle grazing in a typical wood pasture might have hundreds of different plants to feed on. The leaves of trees, fruit when it's available, as well as a mix of wildflowers and grasses, not only help to keep animals healthy and happy, but they make them taste better too. When fed only on high-energy rye grass and concentrated feedstuffs, often made of soya and other monoculture crops, livestock may grow faster, but they often suffer from a range of ailments and require medication to keep them fit. Despite thousands of years of domestication, livestock still seek out the plants that will supply them with the nutrients and minerals that they need. Wood pastures are one habitat where they can find them.

The wood pasture in Ullswater forms part of the lower-lying farmed habitat between the lake and the open fells above. Much of the wood pasture that might have existed in comparable places at Haweswater would have been submerged by the reservoir, leaving the valley with the sharp transition from water to fell that now defines the character of the area. It's satisfying knowing that our current tree-planting efforts will give leafy shade and shelter to the livestock, wild creatures and people in Haweswater's future, recreating some of what's been lost.

Every valley, every farm is distinct, and each needs its champion. The low-lying southern Lake District fells around Coniston, where John Atkinson and his partner Maria Benjamin farm, are a world apart from the high open fells around Haweswater. Their farm, a stone's throw from Coniston Water, sits in richly wooded terrain, and John often says that he farms a thin strip of land between woods and water.

My first interaction with John was through Twitter. He obviously didn't like the sound of what we were doing at Haweswater, but from what I could see it seemed that our farms and our outlook had a lot in common. Overcoming my nerves, I dropped him a line and arranged to travel south to pay him a visit. I didn't expect that I'd necessarily change his mind, but if I could explain what we were doing at Haweswater face to face, he could at least form his opinion based on the facts. I probably ended up learning far more than John did.

John's family have farmed in the Coniston area for generations, and he's another farmer with an enviable, deep-rooted connection to his place. He's a member of the Federation of Cumbria Commoners, which is where I guess he picked up a dislike for our work at Haweswater, for the Federation has regularly been critical of us in the past.

After a quick brew in the warm hurly-burly of their kitchen, I headed out with John to check his cattle. Driving through the woody winding lanes in his bashed-up 4x4, I was soon disorientated. I rarely visit the South Lakes in the summer when it's usually busy with holidaymakers. I'd come on what turned out to be July's wettest day, and our boots were filled with rain by the day's end, but at least the roads were quieter. The tourists were probably wandering disconsolately around the shops in Ambleside or Bowness-on-Windermere, wishing they'd gone to Spain instead.

The first area we visited looked like it had been lifted straight out of the foreground in John Glover's painting of Ullswater. With the shake of a feedbag, John's native-breed cattle emerged from the shelter of the spreading oak trees, trampling on patches of bracken and working around the brambles that were protecting young trees. The undulating ground was flower rich and colourful. Over the wall, gently sloping hay meadows ran down to a reedy wetland, part of a privately owned nature reserve, which John's cattle also grazed. Reed buntings buzzed over the water, and we chatted with the live-in warden about the return of red kites and ospreys.

After a couple more stops in similarly idyllic locations, we returned to the farmhouse. While John went to check on his pigs, I popped into the farm shop. Maria's creative flair has opened up new opportunities for their farm business. Using milk from their small herd of dairy cows and oils from local wildflowers, Maria produces a range of beautifully fragrant soaps. I bought a selection box as a gift for Becki's birthday and also stocked up on beef and sausages. I don't eat much meat, unless I can source it from farms like John and Maria's, where I know that the animals have lived happy lives and made a positive contribution to their environment. The rump steaks that we later ate on Becki's birthday were about the tastiest either of us can remember eating; we were still talking about them for weeks afterwards.

My shopping done, I climbed back into John's truck and we

headed up a steep winding track through open woodland towards Low Parkamoor, an isolated cottage overlooking the *Swallows and Amazons* landscape of Coniston Water and its encircling fells. On that wet and wild day, the farmhouse was an eerily evocative place. Abandoned in the 1930s, it sat empty for decades, until John and Maria found a new use for it. With spartan period furnishings, composting loo and no electricity, phone or internet, Low Parkamoor must make for a uniquely immersive stay – more time-warp than holiday cottage. I helped John shift a couple of wardrobes down the stairs, tying them into the back of the truck with baler twine, and then we walked out onto the fell behind to continue our drenching.

Through the murk, surprising bursts of vegetation started to appear. Vibrant bushes of eared and grey willow were growing in wet flushes, alongside bog asphodel, valerian and devil's-bit scabious. Bog myrtle, one of the upland's most fragrant plants, and said to be a natural deterrent to midges, was flourishing at the bog edges. Bog myrtle is curiously absent from the fells around Haweswater, though there is a layby full of it on the road between Shap and Kendal, four miles to the east of

Bog Myrtle

Swindale, suggesting that it used to be more widespread. It's a species that is well worth encouraging. Thanks to nodules on its roots that contain nitrogen-fixing bacteria, bog myrtle enhances the quality of the soil in which it grows, engineering its ecosystem to the benefit of other species.

We followed a soggy cattle path along the slope of the hill, past rowan and birch saplings growing through regenerating clumps of bell heather, bilberry and ling. A scattering of junipers grew between boulders covered in thyme and bedstraw. With the rain, the whole scene felt thoroughly Scottish.

John told me that the vegetation had been getting richer and more diverse year on year. The thing that surprised me most, though, was that he'd been grazing his sheep up here, as well as cattle. My experience in the fells so far had told me that restoring upland vegetation while sheep were present was close to impossible, yet John's fell had a whole suite of palatable species that were thriving. His secret, he told me, was twofold. The cattle were instrumental, it seemed, and the plant life started recovering as soon as they were introduced. Secondly, the sheep only grazed here in pulses – moved up for a couple of weeks, then removed for a month, then back again for a while, and so on. By mixing it up like this, John kept his animals well-tended, while also leaving space for the rest of nature to thrive.

This isn't to say that the same approach would work everywhere. One of the distinctive things about the landscape around Coniston is how low it is. John's farm is only 50 metres above sea level. The fell behind Low Parkamoor that we walked across went up to about 280 metres, roughly the same altitude as the bottom of the valley in Swindale. Altitude makes a big difference to how resilient habitats are to grazing. The lower the altitude, the longer the growing season, the faster the growth, and the more grazing the plant life can tolerate.

Diversity is the key principle that underpins John and Maria's farm. They keep a range of different breeds of sheep, cattle, pigs, turkeys and chickens. As well as the land that they own, John and Maria also graze for other landowners, including the National Trust. As well as the soap, they make tweed and bags out of Herdwick sheep fleeces that they buy from James Rebanks. They try to sell as much of their produce as locally as possible, keeping money in the community. As well as the cottage at Low Parkamoor, they have another small holiday cottage at the farm, a bunk barn and a campsite. To keep it all running smoothly, they employ several local people. With such a broad portfolio of activities, they are spreading their risks. If the market for lamb suddenly crashes, they can fall back on soap. They are constantly innovating, coming up with new ideas, finding ways to sustain a life in their beautiful surroundings.

Spending time both with the Lake District's farmers and with conservationists made me realize that we all have far more in common than divides us. The desire to become a farmer or the manager of a nature reserve is driven by the same things: a love of the outdoors, passion for a place to which we are connected, and the desire to shape and steward it.

Great Mell Fell, a curiously round hill resembling a balding pate that dominates the skyline from my front door, is the most obvious landmark in Matterdale. Great Mell Fell is one of a pair, with Little Mell Fell just to the east. 'Mell' comes from the British Celtic and means 'bald'. Little Mell Fell still suits the name, but Great Mell Fell is one of the woodiest hills in the area and getting woodier.

Until not long ago, sheep grazed on Great Mell Fell. Without fencing between the wooded lower sections of the hill, and the more open ground further up, they could roam freely, and were preventing the growth of new trees, risking the long-term survival of the wood. Great Mell Fell is an outlying part of Glencoyne Farm, run by Sam and Can. A few years ago, in order to improve the hill's heaths and woods, the sheep were removed and half a dozen ponies took their place.

Great Mell Fell is a rewarding walk, providing breathtaking views in return for modest effort. Becki and I have been dragging the kids up it since they were tiny, and we've watched how its habitats have changed over the years. Even five years ago the land around the summit cairn was short and grassy. Now it's a carpet of heather, bilberry and crowberry interspersed with undulating mossy hummocks, starting to resemble the Atlantic heath I'd seen above Whinlatter. On the last few hundred yards as you approach the gentle summit, thousands of self-seeded rowan trees are emerging.

Down in the woodland, the relaxation of grazing also shows, with plenty of new recruits and a developing shrubby layer under the old ash, oak and alder. Alpine enchanter's nightshade, surely the UK's most mythic-sounding wildflower, grows under the shade of the trees. An unassuming species, with delicate white flowers held on tall flowering spikes, it has more serrated leaves than its commoner non-alpine relative. Away from Mell Fell, alpine enchanter's nightshade is only found in one or two other places in Cumbria, one being around Haweswater – I found sheets of it coping admirably under the bracken on the reservoir's western shore a few years ago where it hadn't been recorded previously. A satisfying new discovery, keeping my plant-hunting obsession topped up.

The National Trust, which owns the hill, doesn't like

portraying it in these terms, but it would be easy to describe what's happening on Great Mell Fell as rewilding. The use of ponies, a native grazing animal, at such a low density, is promoting the rapid recovery of the hill's hitherto suppressed habitats. It's a largely hands-off approach, letting nature do the repairs herself. And it's working.

Looking after land is not a binary choice between farming and rewilding. Matterdale is a mosaic of hay meadows, hedges, more intensively grazed pastures, silage fields, in-field trees, woodland, fenced-out stream corridors, ponds, rushy wetlands, sheep-grazed fells and a wild pony-grazed hill. Each one plays its part. The way that we live with the land will vary from valley to valley, from farm to farm, but we know that a diverse Lake District will be a rich Lake District.

PART 3

Intervention

Life is divided into three terms – that which was, which is, and which will be. Let us learn from the past to profit by the present, and from the present, to live better in the future.

William Wordsworth (attrib.)

CHAPTER 13

Flow

MOSEDALE:
The valley with a bog
(Old Norse)

Of all our land at Haweswater, Swindale is where our efforts are most conspicuous. Rising in the bogs of Mosedale – the wide, lonely valley lying to the south – it is Swindale's glittering beck that has been the main focus of our attentions. From the crashing falls at the valley head, to the quiet stream wandering through woods, meadows and mires, Swindale Beck is a river transformed.

Swindale has a climate all its own. Lying in line with the prevailing wind, the valley has an uncanny ability to draw in bad weather. It's frequently much windier and wetter in Swindale than in neighbouring Naddle Valley, where my cosy office is. One of the most unpleasant nights of my life was spent in Swindale for the RSPB's Big Wild Sleepout, a national event giving families the chance to spend a night under canvas on our nature reserves. Following an idyllic day exploring the valley and its wildflowers, swimming below the falls, and chasing lizards and butterflies, six families and I basked in the warm glow of a wholesome day spent in a spectacular place. As the sun started to sink, we set out in search of nocturnal wildlife, a strengthening breeze waving branches in our torch beams. When the wind became too strong even for the bats that we were watching, we were forced to retreat to the tents. As the torches were turned off, the sound of children chattering was replaced by the moan of the wind ripping through the valley, the squeak of scots pine swaying in the spinney, and the ominous cracking of twigs.

Within less than an hour we were in the middle of a full-blown storm and the next thing I knew, I had a face full of damp polyester. One of the huge hooped poles had snapped, collapsing my tent on top of me. I wriggled out. It was raining now too, sheets of water coming in horizontally, soaking me to the core. The tent was a lost cause. Fortuitously, I had brought a spare, and briefly congratulated myself on such sound planning and preparation. But putting up a tent in a gale that you can barely stand up straight in isn't easy. As I unzipped it, the tent's bag flew off into the darkness, never to be seen again. The wind kept tearing the fabric from my hands, but after what seemed like an endless soggy wrestling match, the cursed thing was up and I dejectedly crawled in.

I felt that this sort of experience deserved to be shared, so despite the hour I called Becki, relaying my miserable endeavours over the sound of the rain slapping on the flaccid flysheet. If this had been a solo camping trip, I'd have packed up and gone home hours ago, but, being responsible for the rest of the group, I had no choice but to tough it out. Almost as soon as I lay down, tent two's spindly fibreglass poles also gave up the fight. My curses disappeared in the gale.

Crawling out again, I noticed the loo tent had also blown over. I didn't want to think about the bucket inside, I had enough on my mind already. I wasn't sure what to do. I couldn't leave the campers for fear that they might be left disoriented in the dark if their tents also collapsed. Walking around them, most seemed to be surviving the onslaught. The sound of snoring from one did nothing to lift my mood. Then I heard the opening of a zip and a mother and daughter emerged into the night. Their tent had started to let in water, so they were off to sleep in their camper van. In the absence of anywhere else to go, I crawled into their abandoned tent's imperfect shelter. One half was still mostly dry, so for the third and final time I settled down to try and sleep.

Morning seemed to arrive immediately and I was still on duty, so with bleary eyes I went to put the kettle on for everyone's breakfast. The tempest had blown itself out, and a gentle drizzle was falling. 'What storm?' the valley mocked. One by one the families arose, looking considerably better rested than I was and, to my slight disappointment, were oblivious to my nocturnal misfortunes. They'd all noticed the wind in the night but, apart from the folks who had decamped to their van, everyone had slept through. As I waved them off later that morning, they all asked if we'd be running the event again next year. I told them I'd think about it.

As I was hanging up the sodden tents to dry back at the office, Laura, one of our residential volunteers, came out to ask how the night had gone. She was a bit puzzled by my account, saying that there'd hardly been any wind in Naddle Valley. The rings under my eyes were probably enough to convince her that it wasn't a tall tale. It seemed that the night's storm had been a special treat just for Swindale.

It could have been worse; at least the campsite hadn't flooded. Mosedale, which sits above Swindale, is one of the loneliest and wettest places in the Lake District. The water from 900 hectares of its boggy expanse drains into narrow Swindale, meaning that flooding is, and will always be, a regular occurrence, especially but not exclusively in the winter months.

≪❧

Water is the essence of Haweswater. Its abundance is what drew the Manchester Corporation to create the reservoir, counting on the reliability of the rain to supply their urban customers. Swindale is part of that supply, with a drinking-water intake halfway down the valley that diverts water via a subterranean pipe into the reservoir.

The Lake District is as famed for its rain as for its scenery, and each affects the other. The rising and falling, cooling and warming effect that mountains have on the air causes more rain to fall than in flatter parts of the world. The rain makes the place look green and lush, gives life to the rivers, and keeps the lakes and tarns topped up.

The climate has a big impact on the plant life too. Epiphytes, plants that grow on other plants, thrive here, obtaining some of the moisture they need from the air, rather than from roots in the ground. Where the remains of these mosses, lichens, ferns and other plants combine with leaf mould in nooks between branches up in the canopy, soil can start to form. Now and then, nuts, seeds and berries find their way into these soil patches, blown in by the wind, in bird droppings, or stashed by squirrels and jays. These can develop into tiny aerial gardens, which often include 'air trees', indicators of temperate rainforest. Air trees are often small, their size limited by the depth of their pocket of soil. But now and again, where they can burrow roots into the rot holes of their host, they can attain proper

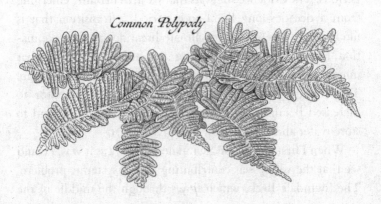
Common Polypody

tree proportions. Because of ash's wide, spreading form, fissured bark and tendency for decay, air trees grow most commonly between its broad branches. I've seen ash cradling yew, juniper, oak, rowan, bird cherry, hawthorn and holly air trees, along with bilberry, bramble and many species of wildflower.

For all its many natural benefits, the damp climate is not universally popular. Lake District farmers have been battling it since time immemorial, and it isn't a battle that's going to be won. Generations of effort have been spent in trying to get water off farmland as quickly as possible. Most Cumbrian farms have complex systems that have evolved over centuries, involving underground drains, ditches, and straightened and deepened watercourses. The landscape has been optimized to transport water downhill as quickly as possible.

Unfortunately, downhill is where all the people live. Cumbria has suffered many serious flooding incidents in recent years, and they seem to be increasing in frequency and severity. Studies of lake sediments have shown that over thousands of years we've oscillated between flood-rich and flood-poor periods. The evidence suggests that we are currently emerging from a decades-long flood-poor period, a transition that is likely being accelerated by climate breakdown. Our population has grown rapidly during this time, agriculture has intensified, and we've built over vast swathes of floodplain. The changing climate is certainly a major contributor to increased flooding events, but the reduced ability of land to store water and slow its flow also plays a part.

When I first set eyes on Swindale, as lovely as it was, I could see that the valley was contributing to this systemic problem. The Swindale Beck, which flows through the middle of the valley, ran as straight as a ruler for the best part of a mile, in a course that was obviously designed to speed the water's flow. Uniformly seventeen feet wide and two feet deep, most of its canal-like route was treeless, its banks raised into levees. To most people, it probably wouldn't have stood out as being a problem. There are few watercourses in the UK that haven't been messed with in one way or another, and so man-made modifications are easily overlooked. A handy rule of thumb that I learned from Tristan Gooley's *How to Read Water* is that 'no natural river or stream will run straight for longer than 10 times its own width. If you see one that does, it's a sign of human engineering.' Keep your eyes peeled next time you're in the countryside; you'll soon start seeing plenty of watercourses that have had their bends pulled out. Swindale Beck was one of them.

☙

Swindale used to be more populated than it is today. Historically, it had at least eleven dwellings as well as a small school and chapel. Today, it has three permanent residences, several virtually redundant stone barns and a load of ruins. There is a tendency to romanticize small isolated settlements like Swindale, but living here would have been a gruelling annual battle for survival. It's high, it's wet, it's windy and it floods; in short, Swindale isn't exactly prime land for growing things, but the people who called the valley home had little choice. Nipping out to buy more hay for the animals if it ran out wasn't an option, especially not in the grip of a hard winter.

To increase their odds of making it work, the people of Swindale, over the course of centuries, modified the valley to suit their needs. Just like in the St Ursus meadows in the Italian Alps, centuries of toil slowly transformed a valley of tall wildflower fens, wet woodland, heaths and tangles of scrub and trees into a mosaic of pasture, meadow and other simpler, more agriculturally productive habitats. Where the soils were deep enough, some plots were turned over to the growing of oats, barley, or root vegetables. Some woodland areas were retained, a few with hazel being cut on rotation to produce useful small poles. Stands of oak would have been used to feed pigs, which would have gorged on their acorns in the autumn, earning the valley its name.

The hay meadows were the most valued feature in this modified landscape, providing the feed to see livestock through the long winters. At some point at least two hundred years ago, the people of the valley straightened the beck in order to protect the meadows from flooding. They did an incredibly thorough job. Both banks were armoured with huge boulders, and the whole reach was deepened and widened to create a neat, square channel without kinks or obstructions through which water could shoot at maximum speed. Getting the water away from the precious hay meadows faster reduced the chances of summer floods destroying the crop.

The generations of work that went into incrementally modifying Swindale deserve respect. The valley's people, just like those of all the Lakeland dales, were industrious and ingenious and they changed their landscape to better meet their needs. Everything they did was right for their time, and they had little choice but to do it. I like to think that everything that we've done since is continuing in that tradition. My colleagues and I are still modifying the landscape to meet our needs, and those of society. The methods are a bit different, and we're now thinking about society in a slightly broader sense, but the rationale is exactly the same.

How best to restore some of Swindale Beck's natural function needed serious thought. A long section ran through the best of the meadows. We needed to make sure that whatever we did contributed to our overall aim of demonstrating that nature conservation and hill farming could go hand in hand.

In true NGO style, we spent time and money commissioning reports and designs. Drones were flown to create a digital terrain model of the landscape, showing us where historic channels were still visible. The Environment Agency's geomorphologists, experts in river form and function, did calculations and ran simulations to determine the optimal course for the restored river. Funds were raised, consents were secured, tenders were put out and quotes from contractors were received. Eventually, by November 2015, everything was in place, ready to start the work the following spring.

Then, in December, Storm Desmond smashed onto the scene. The biggest of a series of storms that winter, Desmond broke all sorts of meteorological records, dropping 341 millimetres of rain onto the Lake District's central fells in the space of twenty-four hours. Carnage ensued. A section of the main road linking Keswick and Ambleside collapsed and took five months to repair. Landslides blocked roads and left scars on the fells still visible today.

Pooley Bridge, which spans the River Eamont as it flows out of the northern end of Ullswater, was washed away entirely, having stood there since 1764. The new bridge wasn't completed until 2020, and other bridges that suffered damage may never reopen. Floodwater inundated thousands of homes in Kendal, Appleby and particularly Carlisle, which, being close to the mouth of the River Eden, bore the concentrated fury of thousands of square miles of river catchment. Farms weren't spared and the flood brought down trees, trashed fields and fences and swept away animals, some of which were found washed up many miles downstream in the days and weeks that followed. The cost of the damage caused by Storm Desmond was estimated at more than £500 million.

Rural communities often show their true colours in times of crisis, and there was a heroic response to support people affected. Temporary accommodation was found for those flooded out of their homes. Volunteers got together to clear debris, working alongside the overwhelmed emergency services and statutory agencies.

After the initial clear-up came the questions. How could we have protected ourselves better? What can we do to stop this sort of catastrophe occurring again? With incredible speed, flood action groups were assembled, and both the local and national government whirred into action to address the problem and to seek solutions. In the search to find something to blame for the disaster, many eyes turned to the hills. Had the upper catchments been more wooded, or if there had been thicker vegetation, perhaps surface water would have been slowed, delaying the flood peaks and reducing the impacts lower down. A light was shone on rivers that had been disconnected from their floodplains in order to protect farmland at the expense of communities downstream. Research was commissioned, funding was provided, and the concept of Natural Flood Management gained traction.

Up at Haweswater, we knew that restoring more complex habitats on the fells, and more natural patterns of water flow, would have a whole range of benefits, including reducing flood risk. Our river-restoration project in Swindale began to feel like our duty. I couldn't wait to get started.

In the first week of May 2016, in fine spring weather, a tractor and low loader trailer shuttled back and forth with two excavators, two tracked dumpers, a small welfare unit, a whole load of bog mats and other assorted engineering paraphernalia. Successfully negotiating the stone walls of the winding lane, they reached the compound that would be the centre of operations for the next few months. I'd worked with these contractors before and I knew we were in good hands. They had done the heavy lifting on a raised bog restoration project I'd managed while working for Cumbria Wildlife Trust. Paul and his dad, Ray, could operate their machines like they were extensions of their limbs. Luke, the foreman for the job, stayed remarkably sanguine, despite some of the challenges we were to encounter together.

The basic concept of our river restoration was a simple one. We had a design, drawn up by the geomorphologists, for a new sinuous route through the valley that was 180 metres longer than the straight, canal-like one with which we'd started. Natural unmodified rivers are dynamic, and as their beds build up over time they periodically shift their position. Over the span of geological time, Swindale Beck had at some point or other probably occupied every part of the flat valley bottom. The job of our contractors therefore was to dig a simple eight-metre-wide trench along the new route, removing the soil and surface vegetation to expose the river gravel below. There was no need to create pools, riffles and bars. Once connected, we were assured that the water, contorted by the beck's new meanders, would form these features for itself.

Like any job involving digging and water, it didn't go entirely according to plan. We hadn't anticipated quite how much effort the previous residents in Swindale had put into drainage. Every few yards the digger buckets severed yet another clay tile, stone or plastic drain, which would gush water into the hole that had just been dug. The weather started fair and stayed that way for all of three days. Then it seemed as though we'd returned to winter, with almost constant rain, and even a period of snow, which at least had the benefit of hiding all the mud for a while.

Even with diggers, changing the course of a river is not an easy business. As we toiled, my respect grew for the people who straightened it with only basic tools, ponies and their bare hands. After a couple of months of hard graft, involving plenty of innovative digger work, problem solving, pump borrowing, silt trapping, mud wallowing and creative thinking, the end was almost in sight when we ran into another major stumbling block.

The downstream end of the original design for the restored route of the beck followed one of several 'paleochannels', the remains of a former route the river had taken through the valley, still visible as a shallow depression in the earth. It was only when the design was being marked out on the ground that it became apparent this particular section obviously hadn't ever been the course of the main beck, but was in fact a channel bringing water from the valley side into where the main beck used to be. This was an embarrassing mistake. We'd trusted too much to the computer models when we should have spent more time observing with our own eyes how water flowed through the valley.

Some people may describe my colleagues and me as 'bookish'; we certainly spend hours in our offices reading reports. I am not sure that I'll ever catch up on the knowledge that farmers have gained from generations of learning about land through observation and practice. As the team problem-solved the design flaws for the new channel, I was left with the familiar feeling of knowing both lots and nothing at all. This was a sharp reminder to keep my eyes on the land.

The designs were changed at the eleventh hour, routing the downstream end of the beck's new course through a lower-lying, wet area, which was clearly the more sensible location. For the rest of the route, trial pits had been dug to check that gravel, which would form the bed of the new course, lay below. However, for this new section of the design, due to the limited time available, these trial pits weren't dug. When the contractors began to excavate the new channel through this section, they didn't find any gravel. They dug deeper, and still there was none. They carried on, all of us expecting that gravel would appear sooner or later. By the time they got to the end of the section, we had a 100-metre-long muddy hole, roughly a tenth of the whole restored length, full of a silty, soupy mess that had several times temporarily swallowed the tracked dumpers.

This was not good. The Environment Agency got nervous. They didn't feel they could allow us to divert the water from the section upstream into the new channel for fear of it washing all this fine sediment away, potentially causing problems for the spawning gravels, freshwater pearl mussels and other wildlife in the river downstream. There was a lot at stake. The work had already cost a little over £200,000, money that had been contributed by the Environment Agency, Natural England, United Utilities and the RSPB. If we couldn't complete the project by connecting up the new channel, all that money would have been wasted, and we didn't even have the funds to be able to return the valley to how it was before.

After some head scratching, the geomorphologists who were advising us came up with a solution. If we made this lower, currently silty section of the beck wider, the water would flow through it more slowly, encouraging deposition of material into it, as opposed to scouring from it. I found my GCSE-level geography and physics coming in handy. The solution appeased the Environment Agency. It was accepted that there would be some contribution of silt to the beck downstream, but we all agreed that the benefits of restoring the beck would outweigh any temporary negative impacts. We had the green light. We were almost there.

The next step was to call in the electrofishing team. Looking a little like ghostbusters in waders, with backpacks and electrified wands, their job was to remove fish from the old straight channel, which would soon be disconnected. Fish are sensitive to electricity, and by passing a low electrical current through the water they can be temporarily stunned, causing them to float up to the surface where they can be whisked up into a net. Because they are so well camouflaged, their cryptic colouration making it difficult to see them from above, far more fish floated to the surface than I was expecting. Salmon, trout, minnows and a couple of lithe black eels were netted, transferred into buckets, and then taken to sections of the beck away from the work area. Mesh barriers were installed at the up- and downstream ends of the straightened channel, preventing the rescued fish from swimming back in.

The long-awaited connection day fell on a Friday. In a carefully choreographed manoeuvre, sandbags were moved to open the upstream gate to the new channel, while we closed off the old one. As the water level dropped, armed with buckets and nets, a team of us, including my gleeful children, leapt into the old drying channel to rescue any fish that had been missed by the electrofishing team.

Watching the water enter the new channel for the first time was exhilarating. As soon as it passed, gravels buried for centuries started to shine on the new riverbed as they were cleaned by the flow. As the problematic downstream end of the restored route started to fill, we were reassured that widening it had been a good move – the water flowed lazily through it, without picking up the silt as it passed. For those of us who had laboured over the past months, losing sleep thinking of solutions to all manner of unanticipated obstacles, this was a momentous and moving moment. Conditions were now set for nature to start to shape the river by herself, returning the detail and complexity that had been robbed by human straightening. We went home for the day, pleased with our work, and looking forward to the final stages of the job that were to follow.

☙

I awoke on Saturday to heavy rain. As the morning progressed, it got heavier, and then heavier still. I knew that if it was raining this much at home, I could count on it being even worse in Swindale. The more it rained, the more I panicked. Just as I was about to head over to see what was happening, the phone rang. George, the project's main geomorphologist, as worried as I was, had already visited the site and his report wasn't good. The beck had already burst its banks, and the flood was rising towards the area where we had piled the soil that had been dug to create the new channel. I rang up our contractor, thinking that we might be able to do some emergency work, perhaps by creating an earth bund around the pile. We arrived at about the same time, diverting our vehicles around flooded roads and battling the strengthening wind. As we drove up to the site, the beck where it ran alongside Swindale Lane was like I'd never seen it before. A colossal, raging torrent was crashing way over the top of the footbridge downstream of our work area and was beginning to cover the road. Compared to this, the storm that blew down my tents two summers before was just a squall.

When we reached the restoration site, there was no sign of our new channel, or the old one for that matter, or the meadows either side. The whole floodplain was entirely submerged. Diggers stood like islands, surrounded by water. Getting to them would have been a challenge, let alone operating them in such a depth. Water was already lapping at the bottom of the soil pile; it was clear that nothing could be done. Concerned that if we spent much more time in the valley we might be cut off and not be able to get home again, we resigned ourselves to the fact that nature had won this episode. We'd just have to wait until the water receded to see what the impact would be.

The rain kept up, and was still falling on Sunday morning, and I knew that there wouldn't be much to see on site. I was unsettled and snappish all day. It was hard to see how this could mean anything but disaster for the beck, and humiliation for me. By evening, the storm started to abate, but I knew it would be hours before the water level dropped. I went to bed restless, plagued with nightmares of what I might find the next day.

I woke at dawn and headed straight for Swindale. My colleague Bill, and John from United Utilities, who had invested as much time and energy on this project as I had, obviously had the same idea, and so we walked over together to survey what we could only imagine would be carnage.

Nothing could have been further from the truth. Aside from a strandline of vegetation indicating the level of the weekend's flood, and a few bits of timber and other debris in some strange places, there was no sign that the valley had been underwater just a few hours earlier. The sun shone; it was a perfect summer's day.

As we approached the downstream end of the restored section of the beck, which had been a silty hole in the ground two days earlier, we were greeted by a scene that took our breath away. The August sun shone through a crystal-clear channel of water, flowing over shining, pristine gravel. A 30-metre-long bar ran alongside the flow, several inches proud of the water. The silt that had caused us such headaches over the past weeks was nowhere to be seen. Carried down from upstream by the surging flow, thousands of tonnes of gravel had been on the move over the course of the weekend's storm, and the river had deposited it all exactly where it wanted it, creating new riffles, bars, pools and shallows and a river through which water moved far more slowly. It was as if a switch labelled 'natural processes' had been flicked; nature was now back in charge.

John, Bill and I wandered around on the new gravel streambed, which was not yet fully formed and still slightly spongy, completely dumbstruck. We had over fifty years of nature conservation experience between us, but we had never seen anything like this before. Our relief was immense, and we cavorted around on the gravel, sinking our fingers into its newly minted surface. I drank the cool water from my cupped hands, the first to taste the crisp flavour of our reborn river.

The geomorphologists had assured us that our problem section of river would undoubtedly fill with gravel as a result of having made it wider, but that the process might take years, or decades even. As it turned out, it happened in under forty-eight hours.

So, what happened to all that silt? Had we caused a pollution problem downstream? During any large storm event, like the one that blew through that August weekend, the levels of silt in any river, technically referred to as turbidity, are always elevated. In a river system as huge as the Eden, of which the Swindale Beck is one of many tributaries, the silt that was washed down that weekend would have been a tiny part of the whole. Monitoring the riverbed at several points downstream after our work was complete showed that everything had returned to normal within a few weeks.

The re-meandering of our beck won't stop Carlisle from flooding, but alongside the other work we've done in the valley, we've definitely slowed the flow. The removal of levees and rock-armoured banks along several hundred metres of beck up and downstream of the restored section lets water access the floodplain more regularly. Without these banks, water can now easily get back into the beck again as the flood waters recede; stagnant pools no longer degrade the agricultural and botanical value of the meadows either side.

This project was never solely about flood-risk reduction. Because the restored channel is so much more curvaceous than before, water flows through it in a very different way. Over gravelly shallows, it ripples and ruffles, drawing oxygen into the water's flow. Where it runs against the banks, it can eddy and cut, winning gravel that can be shaped into new bars and other features further down. Over deeper sections, it slows and, with less energy, drops any solid material that it might be carrying. Riffles, bars, pools and islands now come and go, shunted around with every storm flow, providing a constantly changing habitat for a massive range of life.

Three months after the machines packed up and left Swindale, I was walking along the rewiggled beck on a chilly autumn morning, the frosted meadow grass crunching under my feet. The new channel was bedding in nicely. Every time I visited, there was some new development, a deeper pool here, a little island there. There'd been some bank cutting on one side, but on the other, material had settled out and was already being colonized by plants. I marvelled at how much noisier the new beck was, its water bubbling over rapids, swirling over the bars. The river in its old channel used to flow in silence – few rocks broke its uniform surface.

Suddenly a sleek silver shape shot past me, creating a deep bow wave, then a frantic splash as it passed through some shallows. Atlantic salmon had always been able to get up here, using the fish pass over the water intake just downstream, but seeing them in our restored section for the first time was a thrill. A bit further up, just above one of the recently formed rapids, there was a large circle of gravel that looked cleaner than the material surrounding it. A hen salmon, perhaps the one I'd just seen, had been here, lying on her side, using her muscular tail and metre-long body to sweep the gravel away so as to create a shallow depression, known as a redd. With an attendant cock salmon fertilizing them, the hen had released hundreds of orange eggs into the redd, and then covered them over again.

Like too many other species, Atlantic salmon are not doing well in the UK for a whole range of reasons, one of which is a lack of good spawning habitat. In our old, straightened beck, the torrent had stripped all the smaller gravels from the beck's bed, firing them off downstream, leaving only large, immobile rocks – a stunted habitat of no use to the salmon. By handing the reins back to nature, we'd enabled the creation of a watercourse that salmon could thrive in. I saw plenty more redds that day, and each year their number grows.

I'd always hoped that salmon would benefit from our restoration work, but I hadn't thought they would respond so quickly. Working in conservation, you rarely get instant results. If you plant trees, you can expect to wait half a century before they grow into anything resembling ecologically functional woodland. Restoring a bog, grassland or wetland always involves a period of development where the plants reassert themselves. Rivers, it seems, can recover in a weekend.

Mosedale, where the streams that feed into Swindale Beck originate, has seen its fair share of restoration too. Production-focused grants in the 1970s and 1980s incentivized the cutting of drains into peat bogs in order to improve them for grazing. The grants were popular and precious few upland peat bogs are fully intact as a result. In Mosedale, the poker-straight drains are visible from miles away, incongruous in an otherwise curvy landscape. A couple of feet deep, and a foot or so wide, these ugly, artificial channels reduced the bog's spongelike properties, speeding the flow of the water downstream towards Swindale. As it dries, peat becomes crumbly and is easily eroded, staining the water the colour of black tea. This 'water colour' must be dealt with through a costly purification process before it can be supplied to United Utilities customers. Frustratingly, this state-funded ecological vandalism didn't even succeed in improving the grazing, as was intended. Most of the upland peat bogs I know became dominated by rushes and coarse grasses, which are just as unpalatable and lacking in nutrients as the sphagnum mosses they replaced.

During my first couple of months in the job, John organized a trip to the Watchgate treatment works near Kendal, where the raw water that flows out of Haweswater is dealt with. Watchgate is the largest treatment works that United Utilities operate and it's an impressive and complicated place, an Escher drawing of pipes, screens, tanks and gantries. We were shown the various processes and treatments that were needed to bring the water up to the standard required for drinking. At the furthest end of the plant was a house-sized, dark brown pile, fed by a conveyor belt. This, it turned out, was the dissolved peat that had been removed from the water. Thanks to the chemicals used in the extraction process it was strange and powdery stuff, nothing like peat when it's in the ground.

Some 10,000 tonnes of this reconstituted peat are taken away from the plant each year and spread on the local fields as a soil conditioner. Haweswater's catchment is 10,000 hectares. That means every year a tonne of peat is lost per hectare of land, flowing down the pipes to Watchgate. A tonne of soil is a bit less than a metre cubed, so spread out across a hectare that makes a very thin layer indeed, but it adds up, as I saw. United Utilities have other treatment works over in the Pennines that have to handle far higher levels of water colour, and greater volumes of peat, but Haweswater's pile was still larger than any of us wanted to see. Tackling the issue of dissolved peat and water colour was one of the main reasons that United Utilities started paying more attention to what happened on their land, which in a roundabout way led to us taking over the tenancies at Naddle and Swindale. The size of that pile at Watchgate was a gauntlet laid down.

Working with Natural England and Mosedale's tenant farmer, United Utilities carried out major repair work in 2013, employing contractors to block twenty-nine miles of drains, helping the bog to regain its water-filtering, flood-reducing, carbon-capturing functions. Plugs of peat were pushed into the drains at roughly 10-metre intervals. As the water backed up against them, thousands of shallow ponds were created. Over time, these will fill in with moss and new peat will start to grow.

Mosedale now is a marvel. The drain blocking, combined with the removal of sheep grazing, means that the whole soggy place is rapidly recovering its absorbent layer of multicoloured sphagnum, studded with bog flowers. The ponds vibrate with frogs and toads, and dragonflies soar over their heads. Snipe drum, red grouse gargle, and lizards and voles scurry in the hummocks.

A couple of months after our beck restoration work was complete, I was chatting with Matthew and Jack,* one of the local

---

* Not his real name.

farmers who had been working for us on and off. The impacts of Storm Desmond were still very much on people's minds. Roads and bridges were still being repaired and hundreds of people in Carlisle were unable to move back into their flood-damaged homes.

Jack's at the younger end of the hill-farming spectrum, probably about my age. He does his bit for the local community and is active in several local farming and commoners' groups. Although he was always very open about his dislike of what we were doing at Haweswater, we've always been able to chat about it in a civil way and get along.

I was telling Jack how satisfying it was to have contributed to collective efforts to reduce flood risk. Although the impact of our beck restoration would be small by itself, it was proving a talking point, helping to demonstrate to others how projects of this sort could sit comfortably alongside farming. I shared my hope that we'd start to see more river restoration across the Lakes, and I asked whether he thought there was much scope on his farm, or those of his neighbours. I was hoping I could ignite his sense of civic duty and encourage him to take some action that might also contribute to slowing the water flow. In response he said, 'The thing you need to understand is that we don't give a fuck about people in Carlisle. If people are stupid enough to live on a floodplain, then they deserve to get flooded, don't they.' Matthew didn't know where to look.

I was temporarily speechless. To hear such a view expressed so bluntly stunned me. I guess there are people with similarly strong opinions that run counter to mine in every community. I've heard plenty of conservationists and rewilders expressing ill-thought-out, anti-farming views that have made me wince with embarrassment in just the way that Matthew did. While it might have been a genuine reflection of how Jack thought about things, it certainly isn't the mindset of all farmers. Most think deeply about what they do on their land and the effect it has on others.

Water connects all of us – rivers are the conscience of a landscape, accumulating the residues of activities on land, for better or worse. What we do in Swindale has an impact on how the River Eden behaves thirty miles north in Carlisle. Some of the water winding its way through our beck's meanders ends up coming out of taps in Manchester. The salmon that choose to spawn in our dynamic gravels travelled hundreds of miles from their summering grounds in the Atlantic.

No valley is an island, and neither is any farm.

CHAPTER 14

Flowers

BENTY HOWE:
The hill where the bent-grass grows
(Old English/Old Norse)

In a wild, pre-human landscape, wildlife was constantly churned by natural processes. Where large numbers of herbivores congregated, scrub and woodland were opened up, allowing grassland to dominate. If a wolf pack suddenly set up in the area, the herbivores would be forced to find safer grazing, and the grassland would start to revert to woodland again. Floods and fires could strip the land, leaving soil bare for pioneer plants to colonize. A tree falling in a river might divert it into a new course, the resulting waterlogging triggering the development of a bog, reedbed or fen. This constant tussle between animals, plants, the earth and the weather would have ensured all our habitats were present, but not always necessarily in the same places. The loss of a few hectares of species-rich grassland from one spot isn't a worry when you know it is being recreated naturally somewhere else. But this is no longer the dynamic state that nature has the freedom to enjoy in the UK.

As a species we humans have grown skilled at mimicking

natural processes, employing cutting, grazing, tree-felling, flooding or burning to keep habitats the way that we want them, be it as a playing field, a heathland or a coppiced woodland. Our mastery of nature is what has supplied us with food and fibre and is the basis for much of modern civilization. At times, our interventions can be as good for the rest of nature as they are for us. Many habitats – species-rich grasslands, for example – are botanically richer when managed with cutting and grazing than the woodlands that would develop if left unmanaged.

The nation's Site of Special Scientific Interest (SSSI) network represents the best remaining examples of all our habitats and species. They are the jewels in the crown of our natural heritage – but not everybody loves them. A landscape in its most unmolested condition is constantly changing, its habitats developing from one state to another with the process of natural succession. Some fans of rewilding see any attempt to arrest this process as working against nature. But we're part of nature, whether we like it or not. Debates about whether we choose to manage, or not to manage, to engage or to stand back, take up too much of our energy. In the meantime, species and habitats go swirling down the drain.

Without seismic changes in how we manage our countryside, there is little prospect for reinstating the type of dynamism that would mean we'd no longer need to worry about the loss of our protected sites. So we must look after what we have left, and work for the day when these fragments won't be so isolated and vulnerable.

❧

Several of the hay meadows that carpet the valley floor in Swindale are designated as a SSSI, marking them officially as

sites of national importance for nature. Because of this designation, we work with Natural England, the government's
advisor for the natural environment, to ensure that the meadows are properly cared for. Even more so than the RSPB,
Natural England is often perceived as monolithic, corporate
and inflexible – hardly surprising for a government agency
tasked with enforcing rules and regulations and responsible
for spending large sums of public money. But that doesn't
mean the people working for Natural England aren't driven
by passion, striving every day to do the right things for our
wildlife.

I regularly work with many staff at Natural England, but I
spend most time with Jean and Simon. Jean is the responsible
officer for Swindale Meadows SSSI, and Simon, her husband,
has the same role for the other two SSSIs that fall within our
land, and oversees our stewardship agreements. Both are incredibly knowledgeable ecologists, and I've learned much from them
over the years. They have been central to almost everything
we've achieved at Haweswater, and I count them as good friends.

The work that Jean and Simon have put in to protecting
nature in the Lake District over the decades is nothing short of
heroic. As well as having responsibility for SSSIs, their work
with farmers and landowners to design stewardship schemes
has had a positive influence for wildlife on farms and commons covering thousands of hectares. I've singled Jean and
Simon out because they are the officers I know best, but their
many colleagues are equally committed and effective. Their
work isn't getting any easier, though. Despite nature being
under more pressure than it's ever been, the government
slashed Natural England's operating budget by almost two-
thirds in a decade, and its workforce has been decimated as a
result. As for the dedicated staff who remain, they've had to

take on vastly increased workloads and to accept that part of what the organization should be doing simply can't be achieved.

Jean is kind, considered and deeply committed to nature, but in order to do her job and uphold the rules that protect wildlife on behalf of the nation, she must also be tough. Early in her career she was described in a letter of complaint to her boss as the 'granite face of bureaucracy', merely because she'd informed a farmer that he wasn't allowed to drain and plough the wildlife-rich SSSI wetland that he'd just bought. Because she's involved with so many different sites and schemes, Jean has endured ten times as many bruising arguments as I have, and I suspect that at times they've taken a heavy personal toll. It's a small miracle that she's still standing.

When we took over in Swindale in 2011, our meadows were not in great shape, and Jean had officially categorized them as 'unfavourable'. The previous tenant of Swindale apparently wasn't very interested in wildflowers and so, despite legal obligations to manage the meadows in a way that was intended to protect them, he spread too much manure and fertilizer on them, and grazed them too heavily. This is a perennial problem for the management of SSSIs. If they end up in the ownership of someone who just doesn't care about looking after them, it's likely to be a headache, both for them and for Natural England.

I don't mean to knock the previous tenant of Swindale. The fact that he didn't hand the meadows on to us in as pristine a state as we might have liked was not through any malicious intent. He was just farming in the way that was the norm at the time, focusing his energies on food production rather than nature conservation. That the meadows are still there at all is a feather in his cap. He could have easily ploughed them up, destroying them completely.

The meadows in Swindale were originally designated in 1985 as a particularly special upland type. These 'northern' hay meadows are characterized by big chunky flowers such as wood crane's-bill, globeflower and melancholy thistle. These species were still present in our meadows when we took over, but they were few and far between. The valley was down to a single globeflower plant, bravely clinging to one of the banks of the beck as it flowed through the meadow.

A member of the buttercup family, globeflower is my favourite northern flower. Reaching a height of up to half a metre by late summer, the globeflower's lemon-yellow orbs appear sealed. What look like petals are in fact modified sepals, which in most flowers are green, and typically form a cup in which the flower sits. When you gently part the protective yellow sphere, the globeflower's true flower, including a dense crown-like spiral of carpels, the female reproductive structures, is revealed. The flower is so well hidden that it foxes most pollinating insects. Only one group of flies, members of the genus *Chiastocheta*, has worked out how to penetrate the encircling sepals to reach the flower and nectar within. The flies' tactic is to find

Globeflower

and enter the globes as soon as they begin to develop in spring. They then spend almost their entire lives feeding, resting and mating within their confines, laying their eggs on the carpels. So specialized is this trick that the flies and the flowers now depend on each other. Once the fly larvae hatch, they start tucking into the globeflower's seeds. Fortunately for the globeflower, they also tuck into each other, and in any given year only a small proportion of seeds are consumed before the fruits that have formed from the flower's ovaries split and drop their contents, seeds, fly larvae and all, to the ground below. The flies then pupate and spend the winter in the soil, emerging in spring to make their home in the globeflowers once more.

There are fewer than 900 hectares of northern hay meadow left in the UK, virtually all of which is in northern England. To be looking after 2 per cent of the whole national resource of this habitat is a serious responsibility. One slender upside to the catastrophic loss of hay meadows is that there has been a concerted effort to work out how best to restore them. Working with Jean, we started to enact a plan for their recovery.

Hay meadows thrive where soil nutrient levels are low. Where they are too high, grasses can get the upper hand over flowers. Our soils were a bit rich, but that was easily addressed by ending the application of fertilizer and manure. We still had most of the important big, blowsy species present in the meadow, but they weren't as numerous as they should have been. To encourage them to spread, changing the timing of grazing was key. In order to attain their larger size, bulky wildflowers start growing earlier in the year than some of their smaller companions. The previous tenant in Swindale frequently used the meadows for lambing, meaning the meadows were grazed right through to April or May, disproportionately impacting earlier growing species. By lambing elsewhere on the farm, we can now shut up the meadows much earlier in the year, and so the bulkier plants are spreading.

Moving to an earlier shut-up date for our meadows is more in line with local farming traditions, too. At different periods through history, the traditional shut-up date has varied, but I've found several references to it being in February or March. Without the boost of chemical fertilizers, farmers had to rely on the muck created by their livestock and occasional dressings of lime, produced with serious effort in local kilns. The obvious way to ensure a good crop was to allow the hay as much time as possible to grow, so an early shut-up makes total sense. The fact that melancholy thistle, globeflower, wood crane's-bill, knapweed and bistort also benefited from this is a happy accident.

The timing of cutting is also important. We never cut our hay before late July to ensure that the seeds can ripen and drop back into the meadow as the hay is turned to dry it. We try not to stick to a steady state. A later cut in some years means that late-flowering species such as knapweed, devil's-bit scabious and great burnet have a chance to seed before the hay is baled. The bad weather helps to mix things up, and there have been some years where we haven't had the three clear days of dry weather we need for cutting until well into September; and occasionally we've only been able to cut parts of the meadow, relying on our sheep and ponies to graze the hay down instead. Nature loves variation, and I'm sure that these challenging hay-cutting years have done the meadows a power of good.

Further up Swindale, towards where the Hobgrumble Gill waterfalls come tumbling into the valley, we have other undesignated meadows. Because these lacked the protection afforded them by a SSSI badge, their wildflowers had been even less well looked after by the previous tenants. To get these back up to scratch, we harvested seed from the best bits of the

SSSI meadow, and then after gently harrowing to open the soil, we scattered the seed into it. We added a selection of locally grown plug plants to add further diversity. Then all we had to do was sit back and wait for nature to take its course.

Five years later, Swindale's meadows are a joy to behold. Colourful, diverse and buzzing with life, walking through them in late June is one of the most uplifting experiences there can be. Great clumps of melancholy thistle tower over the herbal understorey, a haze of eyebright peering up from below. Patches of bright pink bistort are spreading, and heath spotted orchids have appeared. In late summer the staccato of dried yellow rattle seed heads accompanies every step. There are hundreds of globeflowers now, spreading alongside saw-wort, ragged robin and meadowsweet in the wetter areas. Tinkling flocks of linnets and goldfinches feast on the abundant seeds, jostling with chimney sweeper moths, meadow brown butterflies, bees and hoverflies for airspace.

With the restored river meandering its way through this flowery wonderland, we've got a lot to be proud of in Swindale. What's perhaps most satisfying is that we've done this work while continuing to farm. The crop we take from the hay meadows is an important part of our system, feeding our livestock through the winter. True, we get less bulk of hay per hectare than a silage field, but with reduced numbers of animals we've found a balance. The restored beck spills into the meadows more regularly than it used to, but the water can also return to the channel more easily when the flood subsides. The meadows seem to like this new natural flooding regime, which is contributing to their continued improvement. All in all, it's a powerful demonstration of farming with nature.

I've come to think of our meadows in Swindale as one of our two flower sanctuaries, places where the focus of our management is on enhancing botanical diversity. Our other one is a lot steeper.

Not long after our study tour to Scotland, Simon, John, Dave, Bill and I were sitting at the top of Harter Fell, 750 metres above sea level. The full curve of the reservoir was laid out below us, with Bampton Common rising out of the left shore and Mardale from the right. Patches of this landscape were approaching the richness we'd seen at Carrifran and Ben Lawers, and we were discussing how we could apply what we'd learned in those two places to allow them to expand.

We'd taken a path less travelled before our lunch stop, picking a way through the crag to get our collective fix of Haweswater's richest slice of montane habitat. There are almost certainly discoveries still waiting to be made in these flower-strewn gullies, and so it's always worth going for the scrambly route, even if it might take a little longer. Like some crazed experiment in psychedelic rock gardening, Harter Fell's sheer terrain was exploding with yellow, purple, pink and white, rich with succulent roseroot and primordial ferns, saxifrages, lady's-mantles, juniper, heathers, mosses and lichens. Having seen the botanical response to a release from grazing in Scotland, I could almost sense our exuberant plant life's desperation to recolonize the land below.

The glossy green, saw-edged fronds of holly fern can only be found in a handful of places in England, one of which is Harter Fell. Already pushed into inaccessible corners by grazing, the depredations of the Victorian fern collectors knocked holly fern back even further, resulting in its present-day rarity. Simon, who has been exploring Harter Fell's crags for many more years than I have, found a few new clumps of holly fern in an area where none of us had seen it previously, further confirmation that the colony was the largest for the species in England.

Popping out onto the bare, heavily grazed flat summit of Harter Fell, after the vibrancy of the ungrazeable crags below, is always a shock, a reminder of how much work we still have to do. We walked along the undulating ridge, peering into the tops of the gullies for any other interesting plants, listening out for ring ouzels. We settled ourselves in the lee of some boulders, where a deep cleft protected a hunkering patch of dwarf willow from the sheep.

With the sensation of the exclosures at Ben Lawers fresh in our memories, a simple, almost obvious idea struck us all. Directly below us, separated from the surrounding common land by walls and fences, was the 75-hectare outlying section of Naddle Farm, land which had lost its farmhouse to the water eighty years earlier. It was the perfect place to create a grazing exclosure of our own.

Stretching from the base of Harter Fell to the southern shore of the reservoir, these fields have lots of diversity, with streams, bogs, gravelly flushes, boulders, rocky knolls, screes and slopes. There were some areas of remnant heath, a few bracken patches, mire and lots of species-poor grassland. It was land completely within our control, with minimal value for grazing, so we could fence it and manage it in any way we pleased. With such a huge and diverse range of seeds dropping from the crag above it, and watercourses to deliver them, if we could keep out the sheep and the deer there was a good chance that plants from the crags would rapidly colonize the lower ground. The only thing stopping us was the cost of the fencing.

Dave, who was the RSPB's regional ecologist at the time, is also president of the Alpine Garden Society. I love seeing plants in their natural environment, but I'm not much of a gardener.

Dave on the other hand seems to be able to grow anything, and his garden and greenhouses are a kaleidoscope of alpines from across the globe. Dave thought that our exclosure idea could be right up the Alpine Garden Society's street, helping them to realize their long-standing but unfulfilled aspiration to conserve British alpine plants in the wild.

That evening, he made a few calls, and a site visit for some of the Alpine Garden Society's trustees and staff was organized. Shortly after, we had a shiny new partnership in place and a generous commitment to pay for the fences. The fencing was finished in March 2017, keeping both sheep and deer out of an area the size of about fifty football pitches.

In the Ben Lawers exclosures there had been some planting in the early years, but nature soon made further interventions pointless; once the seed source was re-established, natural regeneration became the driving factor. We're doing the same, and four years after the fencing went in, we are still in that early planting phase, looking forward to when our efforts will be unnecessary. For the time being, it's rewarding and engaging work.

We're not looking to augment our exclosure with species from far afield; we've got no interest in creating a botanic garden. We only plant species from the local area, or those which have been recorded in the area historically. Where we can, we grow plants from seeds collected from the immediate surroundings. There's lots of genetic variation within any species, so growing plants from seeds collected as locally as possible means they stand the best chance of thriving in the local climate.

Many of the Alpine Garden Society's members are expert horticulturists. They help by growing plants at home or with planting days in the exclosure. Many are also rather elderly, and not especially well suited to scrambling around in crags hunting for seeds. This has given me an excuse to develop a new fascination with a stage of plant development that I hadn't really appreciated previously.

⋙

As much as they might also appeal to us, the colour, shape and scent of flowers are really adaptations to attract pollinators. Once pollinated, seeds swell in the ovaries of flowers, each one a tiny miracle with all the instructions and ingredients to carry a plant's genes into the future. The magic of seeds, along with their diverse tactics for dispersal, have an appeal all their own.

In order to avoid competition between a parent and its seedlings, plants have evolved an array of tricks for dispersing their seeds as far away from their parents as they can. Wood and meadow crane's-bill, for instance, have a spring-like mechanism that catapults their dark brown seeds in four different directions. The tiny seeds of ragged robin, globeflower and red campion develop in capsules at the tips of tall flower stems. The lids of these capsules fall off when the seeds are ripe, and as the tall stems are blown around by the wind, the seeds are flung hither and thither. The seeds of species in the pea family, such as gorse, broom, bitter vetch and petty whin, develop in pods, which burst when they dry, flinging their seeds away. The crackling sound you can hear as you walk past a gorse bush on a hot summer's day is the sound of its hairy black pods splitting open. Plants that grow berries tempt animals to swallow their seeds and carry them away in the gut, conveniently adding nutrients as they're defecated. Spiky or sticky seeds like water avens catch a ride in the fur of passing animals. Valerian, willows, thistles and many others have their seeds attached to downy umbrellas, which float away on the breeze, carrying them many miles if the wind is strong enough.

The seed collectors' trick is to visit the plants after the seeds have ripened but before their dispersal mechanisms have been deployed. Go too early, and the seeds might not germinate. Visit too late, and they may all be gone, flung into space, guzzled by wildlife, or shaken into the surrounding soil.

I'll admit, it's becoming another obsession. My mental map of Haweswater is now dotted with all the best places to collect seeds. Because it's harder to find my target species after their flowers have finished, I've got various markers around the place, sticks stuck in fences, or canes in the ground to help me locate my tiny quarry. I never set out in the summer months without a few waxed seed bags in my rucksack, and our kitchen table is often cluttered with plastic tubs of seeds and spent flower heads awaiting sorting. It is captivating to focus so closely. Not only am I more appreciative of the steady seasonal development of flowers from bud to seed, but I'm getting to know some of the other tiny lives that accompany them.

I was out on one of my foraging missions one July morning, and came across a hedge with an abundance of red campion. I was lucky with the timing; their brown capsules had all lost their lids and dark spherical seeds were rattling loosely inside. Red campion grows abundantly on some of Harter Fell's ledges, so it was on my list of species to collect for the exclosure.

I started gently tapping the capsules into my collecting bag, leaving plenty behind to contribute to the next generation in situ. As I worked my way along the hedge, every third capsule or so I looked into was empty, with many of the vacant ones growing from the same stalk as others which were brim-full. All would have been exposed to the same breeze that would have shaken the capsules around, so it was a puzzle as to how some could have been emptied without their neighbour also losing its contents. Pinching the base of another capsule to tip it into my bag, instead of seeds the head of a pudgy brown caterpillar emerged, glaring accusingly. As I took my fingers off, the caterpillar grumpily retreated, tucking itself back inside, its subtle brown stripes blending perfectly with the colour of the capsule's innards. Here was the culprit.

This little thief was the caterpillar of the aptly named campion moth. Richly patterned in brown, cream and purple, the adult moth lays its eggs on the campion flowers, into which the tiny larvae burrow once they hatch. As the seeds develop in the flower's ovaries, the hungry caterpillars find themselves surrounded by their favourite food, which they snack on, clearing the capsule out completely before emerging at night to move on to another one. As the summer progresses, the caterpillars eventually outgrow their capsules and switch to eating the campion leaves before pupating into the adult moth. Having got my eye in, I could spot the caterpillars curled into a perfect nose-to-tail circle, hiding just below the frilled edges of their capsule shelters. There was something beautiful about how perfectly the plant and the insect fit each other, how evolution had entwined them.

I regularly find little creatures in my collecting pots, scooped up alongside the seeds of the plants they're feeding on. The brown oval seeds of meadow crane's-bill often come with tiny orange grubs, grey fly larvae with melancholy thistle and knapweed, spiders and leafhoppers with yellow rattle. There are innumerable intricate, intimate interactions between flowers and insects, more than any human can hope to understand. Knowing that there will always be more to learn is one of the things that makes ecology so engrossing.

The next stage of the process I leave to folks with greener fingers. Each late summer, I post out the small white packets of the seeds I've collected to Dave, Jean and Simon and the local Alpine Garden Society members, who get to work growing them. I like to think about this dispersed nursery operation, small hubs of lovingly tended Haweswater plants in greenhouses and gardens across the county. A year or two later, deliveries of healthy young plants start to arrive. We keep them in our polytunnel for a while, both to check that they haven't come with any freeloading species that we might not want, and to grow them on a bit more. The final step is planting them out into the exclosure, which is usually done by our trusty Haweswater volunteers over the course of a few days in the autumn, when there's a good chance of plenty of rain to water them in.

It's probably not the most efficient way of doing things, but it creates a load of brilliant opportunities for people with different skills to give something of themselves back to nature and to build a lasting connection to a place, like the many volunteers who planted trees at Carrifran. Through these collective, collaborative efforts, we've created something that was missing from my early years at Haweswater – a feeling of community. Sharing a purpose and celebrating successes is essential sustenance for any endeavour.

We've planted plenty of trees too, both in the exclosure and widely across the rest of our land. Spike and Jo are the master growers and keepers of the Haweswater tree nursery. There's been a nursery at Haweswater for decades, but we've recently expanded it, building raised beds on the footprint of a now redundant sheep barn at Naddle Farm. With nuts, seeds, berries and cuttings collected from our woods and fells, with the help of yet more volunteers, the nursery produces thousands of juniper, hazel, rowan, wych elm, holly, oak, hawthorn, birch, aspen and willow saplings every year.

It's somehow easy to forget that trees are flowering plants too. Some of their blossoms, like the tiny pink star-like flowers of hazel, are easily overlooked. Others are more conspicuous, and rather odd, like those of ash which start out resembling unripe blackberries before growing into something more like sprouting broccoli. A blossoming hawthorn must have tens of thousands of delicate white flowers, each one as valuable to a pollinator as any daisy or violet. Between them, our range of native trees and shrubs are in flower almost all year, with hazel starting as early as January. The baton is then handed to aspen, alder and wych elm in February, blackthorn in March, cherry and birch from April, and so on right through to gorse, bramble and ivy finishing the relay in autumn.

Planting trees and wildflowers into the Mardale exclosure is helping to build its diversity, and we're now able to collect seeds from plants that we sowed during the first few years. Because we've kept it local, they could realistically have arrived by themselves. All we're doing is giving nature a boost to speed things along a little.

The planting work has been rewarding, but the most exciting changes in my view are those that have happened by themselves. Without the nibbling of sheep and deer, the plant life is returning at speed. In the areas closest to the crags, early purple orchids and vast sheets of mossy saxifrage have appeared out of nowhere. Devil's-bit scabious, goldenrod, wood crane's-bill, lesser meadow-rue, grass of Parnassus, yellow mountain saxifrage, ferns and willows pick out the stream sides where their seeds have been washed down from above. In the peaty areas, between the growing hummocks of bog mosses, great golden mats of bog asphodel are spreading, their furry yellow flowering spikes brightening the bog. Just like in the meadows, the plants have returned in tandem with the insects. Hoverflies and ichneumon wasps throng the weighty flower heads of angelica and butterflies drift by, drunk on nectar.

There is hope for the plants on our fells. In the right places, a rest from grazing can produce spectacular results in a short space of time. Our ongoing plan for the exclosure is simple and satisfying. Bar a little more planting and keeping the fences in order, the plan is to do nothing. There are few places where nature is given that level of freedom. It wouldn't happen everywhere, but in some places a hands-off approach works wonders.

It took about fifteen years before the exclosure at Ben Lawers was significantly transformed. We're further south, and at a slightly lower altitude, so I'm optimistic that things will happen faster for us. If their experience is anything to go by, the richness inside our fence will only increase as time passes, giving its 60,000 or so annual visitors a multi-sensory experience that is hard to find elsewhere in the Lake District. The sight and smell of flowers, and the sounds of insects and birds, will be inescapable.

In a few years' time, visitors to Mardale will receive the same visceral blow that I had when I first walked through the gate at Ben Lawers. They'll get a glimpse into an alternate reality, a slice of the Lake District where flowers are everywhere.

CHAPTER 15

Fells

BEASTMAN'S CRAG:
The rocky height of the cattle-man
(Old English/Old Norse)

As soon as plant-eating creatures appeared on the planet a few million years ago, their food had to adapt to survive. The range of responses partly explains why we have such spectacular diversity of plants across the globe. Some developed defensive strategies like spines or toxins that made them poisonous or unpalatable. Others plumped for rapid growth or the production of prodigious quantities of seeds, so that even if they were grazed, there was a good chance some would survive through to the next generation. The trees opted for tough protective bark and a size that put their tastier parts out of reach of many grazing mammals. Mosses opted to hide in plain sight, small and prostrate, and not worth the effort needed to get at them. No one method gives complete protection from the herbivores, and every plant must endure their nibbling to some extent.

Insects, with their rapid reproductive rate, are often capable of keeping up with this evolutionary arms race, and for every defensive toxin that a plant species has developed, an insect has found a way around it. Other than bees, we don't tend to farm insects very often, so it's the impact of mammalian grazers – sheep, cows, ponies, pigs and deer – that I'm most concerned with.

One of the most effective defences against grazing is employed by the grasses. Most flowering plants have their growing points at their tips. These are where plant cells divide and grow, so that in a typical flowering plant the youngest cells will be at the top. When a grazing animal comes along and nips this growing point off, the plant's growth is stunted and its hopes for flowering and setting seed that year are dashed. Grasses evolved to work around this, and over the course of aeons their main growing points shifted downwards, to the places where leaves meet stems, giving them a much higher tolerance to grazing. Though it might not seem like a dramatic adaptation, it's had huge ramifications. As a species, we humans have benefited enormously from grass's survival ability. Wheat, barley, rice, sugar cane and maize are all grasses which, together with grass-fed livestock, keep most of us fed.

Natural grasslands occur across the globe in places where the climate is less favourable for the development of woodland and where grazing animals maintain open conditions. The most extensive of these are the American Plains and the Russian Steppe. Our climate doesn't favour natural grassland, and before widespread human intervention, the generally accepted view is that the UK would have been a landscape of woodland, scrub and wetland, with patches of open ground maintained by roaming herds of wild grazers. Grasses were a part of that primeval landscape, but they wouldn't have been as dominant as they are today.

Thanks to the low-oxygen conditions in peat bogs, like the

ones in Mosedale and at the top of Rowantreethwaite Gorge, plant material that falls into them is preserved, providing a window into our botanical past. With painstaking effort, analysis of pollen in soil samples can enable our ancient flora to be reconstructed; as with polar ice, the deeper down you go, the further back into the past you travel. Studies in the Lake District show a massive increase in grass pollen, starting around 3,000 years ago, neatly coinciding with the period that Neolithic people were expanding settled agriculture, clearing land for livestock grazing. As grass pollen increased, so tree and heather pollen declined. This increase in grass would have crowded out many flowers, just as it still does today.

✍

The tug-of-war between flowers and grasses can be neatly demonstrated by tinkering with your lawn. If, like most people in the UK, you cut yours when the grass approaches ankle height, you're tipping the balance in favour of grasses and against the flowers. Each spring, I leave a few islands in our lawn unmown. Given a free hand, I'd leave more, but a ragged, flowery lawn isn't all that compatible with Elliot and Aphra's designs on our outdoor space. So, as the islands grow long and shaggy, they become goalposts or dens in games of tig, surrounded by grass that I grit my teeth and mow. They rapidly sprout cuckooflower and pignut, quickly followed by common sorrel, mouse-ear chickweed, germander speedwell and meadow buttercup. I'm slowly augmenting these mini-meadows, sowing seeds of yellow rattle, oxeye daisy, knapweed and goatsbeard collected from the local verges. The bees and butterflies love these unmown patches, and when I give them their annual cut in late summer, ground beetles, spiders and earwigs scurry in front of the mower.

At the farm scale, this lawn analogy explains why a hay meadow is so different from a pasture. When a pasture is grazed most of the year round, the livestock are constantly mowing, repeatedly nipping off the growing points of the flowers, reducing their prevalence in the sward over time. The grasses handle the pressure much better, thanks to their basal growing points, and so the sward ends up being dominated by them. Seeds can survive in the soil for different lengths of time depending on the species – heather seed for example can stay viable for decades – but none lasts indefinitely. A perennial flowering plant may be able to live for many years in a vegetative, non-flowering state, but if the grazing pressure is maintained every spring and summer, preventing flowering and not allowing seed to develop, the plants that produce flowers will slowly dwindle and disappear altogether.

A traditional hay meadow enables flowers to thrive. Livestock are removed for the spring and early summer months to let the hay crop grow to make it worth cutting. This break from grazing allows the flowers to open, for their nectar to be drunk, for pollen to be exchanged and for seeds to form. Come the harvest, the flower has served its purpose and seeds are helpfully scattered across the meadow by the scythe, the hay-bob or the cutter, topping up the seed bank in readiness for the following year.

If we scale up again, to the level of the landscape, the lessons learned on my lawn are still useful. On a typical hill farm, sheep spend the summer months up on the fell, at precisely the time the flowers are opening. The areas of upland pasture in the Lake District are large, and if the numbers of sheep were anything like they were seventy years ago, and still are today in the Alps and Norway, then plenty of flowers would escape the nibbling, and everyone would be happy. The post-war farming subsidies, which drove sheep numbers to levels where every square inch was grazed, shifted the balance even further in favour of grass. Although sheep numbers have started to drop in recent years, thousands of hectares of the Lake District fells – much of our land at Haweswater included – are still dominated by coarse grasses and only the toughest or most prostrate flowers get the chance to appear.

*

In 2001, a tragedy occurred that kept the mower locked in the shed. An outbreak of foot-and-mouth disease resulted in untold horror for farmers across the UK, leading to the eventual slaughter of over six million sheep, cattle and pigs. Foot and mouth is a highly contagious virus which causes fever, severe blistering, and sometimes death, particularly for young livestock. Due to the speed at which it can spread, the only effective way to get foot and mouth under control is to cull the livestock on infected farms. Cumbria was the epicentre of the crisis, where 44 per cent of the affected farms were found. The trails of smoke caused by the funeral pyres of burning animal carcasses are etched into the memories of Cumbria's rural population.

For a farmer, livestock represents a life's work. The constant tweaks of breeding and selection, purchases and sales, improve the flock or herd over time. Many farmers name their animals, get to know them as individual characters, and strive to give them as wholesome a life as possible. To see all this love and toil literally go up in smoke, without even the justification of feeding people, must have been devastating beyond belief, and it's a trauma that still feels raw.

The countryside was effectively shut down and movements

of animals were banned, including those of sheep up onto the hill pastures, where there would have been a high risk of disease transmission due to the mixing of flocks from different farms. Public access was banned in some areas. In others, special measures were put in place to limit the spread of the disease. During our 2001 summer break from university, towards the end of the crisis, Becki found work disinfecting the boots of walkers heading up Helvellyn. I joined her for a couple of shifts, stationed in a little shed near Glenridding with a tray full of chemicals. The usually buzzing tourist routes were eerily quiet compared with a normal summer, and a palpable gloom shadowed the fells.

Yet in the midst of the disaster, as the hills received a break from grazing that they probably hadn't had for centuries, an unexpected botanical episode was unfolding. Jeremy Roberts, the botanist who introduced me to pyramidal bugle, witnessed it first-hand and wrote:

For thirty years I tramped the high moors of the Cross Fell range. I often used to wonder how the mountain vegetation might respond to a reduction in the severe sheep grazing to which the tops had been routinely subjected for so long. When the fell sheep were lost through foot-and-mouth culling, that chance arose.

By the late summer of 2002, vegetation had had two seasons of untrammelled growth. An astonishing sight was revealed! Over huge areas, the previous monotonous stretches of mat-grass and rushes were now enlivened by many other species reasserting themselves, no longer nibbled down to the ground by the selective appetites of the sheep. In the endless breeze, cotton-sedges made billowing white swathes, tall grass heads swayed gracefully and innumerable mountain flowers starred the turf with white, pink and yellow. Some of the rarest and most significant species, such as marsh saxifrage and alpine fox-tail, flowered in unprecedented profusion. The latter appeared in areas where hitherto it had been overlooked.

The high rills were made pink by sheets of hairy stonecrop and chickweed willowherb, both usually inconspicuous plants. A local botanist found two colonies of a sedge known on Scottish mountains, but never before seen in England. It is clear that these plants contrived to outlast the epoch of over-grazing, most usually by having creeping stems at or below ground level that even the close-nibbling sheep failed to reach. Now, I believe, it is time to cherish what we have left, and perhaps allow these special mountain species to re-take some of the ground they have lost.

Jeremy wrote this almost two decades ago, but the change he hoped for hasn't happened. Farmers were financially compensated for their losses from foot and mouth, the sheep returned, albeit in slightly lower numbers than before, and the flowers went back into hiding. This brief glimpse of the fells without sheep confirmed what many people already suspected: a rest from grazing can allow for a massive recovery of wildflowers. In some places, it really is that simple: remove the grazing and, hey presto, flowers reappear.

Our exclosure at Mardale Head is our largest, but we've also fenced livestock out of many smaller areas, particularly along watercourses. Inside these protective fences, the banks of streams are now filled with meadowsweet, greater bird's-foot trefoil, sneezewort and mint, with willows and alder also start-ing to appear.

In the parts of our land at Haweswater where the terrain is high, steep or very wet, where there is a good seed source or where there is some natural disturbance, like on the edge of a stream, our aim is to minimize the grazing as much as we can. We're already seeing good results, and there's more to come. In the longer term, once restored, many of these areas should be able to sustain a degree of grazing again, but most have a long way to go before they'll be resilient enough.

In other places, simply stopping grazing won't achieve such good results. In habitats where grasses have been given the upper hand by livestock, banishing the animals can result in the grasses taking an even firmer stranglehold. The dense, ground-smothering mat that can result makes it almost impossible for seeds of any species to reach the soil and restore diversity. These species-poor, grass-dominated habitats have got stuck, pushed beyond a point where they can recover without more intervention. The same is true for dense beds of bracken. Even though bracken usually indicates places that are suitable for woodland, if the tree seeds can't reach the soil through the thick bracken litter, they haven't got a hope of reclaiming any ground.

In some places, the right tool to repair the damage caused by overgrazing is, perhaps confusingly, grazing of a different kind. Striking the right balance between too much grazing and too little is the subject of interminable debates in land-management circles.

⟨⟨⟩⟩

In our first few years at Haweswater, we carried on with the same sheep-dominated system we inherited from previous farming tenants. We cut the numbers a bit, and created several exclosures, but overall, despite what some people liked to think, we were still solidly a sheep farm.

Having seen the dramatic response resulting from the switch

from sheep to cattle at Geltsdale, Haweswater's older sister reserve in the North Pennines, we knew that cattle would have been a better choice of grazing animal for us. Unfortunately, we didn't have a free hand. Mainly because of impacts to water quality resulting from intensive dairy farming, United Utilities were cautious about cattle on their catchments. We had little choice but to agree to limit our cattle numbers.

Our one exception was a 30-hectare scruffy wedge of land in Naddle Valley, a mile or so up the track from the office. Its tree-fringed mire was dominated by vegetation too tall and coarse for sheep to tackle. Because the area was part of a SSSI, Natural England insisted that it had to be grazed by cattle in order to sustain its wildlife. Over time, John helped to assure United Utilities, through his involvement with Wild Ennerdale, that low-density, outdoor cattle grazing actually enhances water quality by helping to rough up the land and promote the regeneration of trees and scrub, creating a more diverse landscape which works as a better filter for water. So, we've slowly expanded our cattle-grazed areas.

The contrast across the two sides of the fence that separates the cattle and sheep-grazed fields in Naddle Valley is growing increasingly noticeable. On the cattle side, their regular trampling has seriously weakened the bracken, young alders are popping up along the edges of the streams, and birch and hazel are spreading out of the woodland. The tall vegetation in the mire gets more diverse each year, with dark purple northern marsh orchids growing alongside a host of tall flowers. Sheep would have avoided this area completely because of the poor grazing and saturated ground. The cattle don't spend a huge amount of time in the mire either, but even their occasional trips through it are enough to stop the coarse grasses and rushes from dominating.

I love going up into the valley and checking on our belted Galloway cattle. They are a small breed, dark, with a broad white band around their middle, resembling giant hairy liquorice all-sorts. If it wasn't for their solid frames and semi-wild nature, you'd almost call them cuddly. For animals with a huge white stripe, they're surprisingly good at hiding. I often find them standing calmly in the shade of the valley's many trees. Something deep in their genes must put them at ease there, the legacy of their primitive, free-roaming wild ancestors.

We've put ponies back to work too. Every year, Nicola, a local breeder, lends us four young fell ponies to look after part of the meadows in Swindale. Fell ponies are the Lake District's traditional breed, their thick, dark coats protecting them against our cold and damp climate. They've been used as pack horses since at least Roman times, their sturdy frames, strength and sure-footedness making them an essential component of rural life until they were replaced by cars and tractors.

Nicola and a handful of other enthusiasts are keeping them going through sheer love. I can understand their obsession. With manes often kept long, all the better to dramatically flail in the breeze, fell ponies are patient and gentle, tough but graceful. Although a domesticated breed, they have a wild aura that is dramatic and stirring, living emblems of the wind-swept fells. That's not to say they can't cause problems. In some places their numbers have been allowed to grow to levels which, in combination with the grazing of sheep and deer, has contributed to overgrazing. Like everything, it's a matter of balance. Although her ponies are not as essential to human endeavour as they once were, Nicola has found a new niche for them, lending the animals out to provide a conservation grazing service to land managers like me.

During the 2018 'Beast from the East', one of the most extreme winter storms in living memory, Nicola had some of her ponies grazing up in the Pennines. The heavy snow meant that she couldn't bring them any feed; they were completely cut off for weeks. She knew that they were tough, but not seeing them for such a long period with the hills blanketed in snow, drifting to three metres deep in places, was a real worry for her. The ponies proved themselves even hardier than she had expected, trotting over to her when she was finally able to get through, looking none the worse for wear.

At the margins of our meadows in Swindale there are wet areas where, as in the mire in Naddle Valley, the vegetation is too coarse for sheep. Like cattle, fell ponies will happily eat rough grass and rushes, even in the winter months. What this poor fodder lacks in nutritional value, the ponies make up for in the quantity that they guzzle, knocking it back with impressive speed. Their grazing has been transformational. On one of the wet valley sides above a section of our re-meandered beck, where before there was just dense purple moor-grass and soft rush, the gaps left by the ponies have blossomed with

Devil's Bit Scabious

extraordinary carpets of heath-spotted orchids and devil's-bit scabious.

The ponies can be a bit of a handful. We electric-fence them into the areas we want them to graze, but they like their freedom, and during the few winter months that we enjoy their company they nearly always get out a few times. Spike usually has the job of trying to round them back up again, but they have a belligerent habit of objecting. I've lost count of the times when he's come storming back into the office at the end of some unsuccessful pony wrangling, shedding a dripping jacket, swearing himself blind that he'll never have anything to do with them again. For some unfathomable reason, they seem to like me better. When I tell Spike that I've rounded them up the following day without any bother, it doesn't generally do much to improve his mood.

⋘

Watching the long manes of the ponies rippling in a snowstorm, or the dark curls of the cattle shedding the rain as they huddle beneath the alders, affects me in a way that I didn't think farm animals would. These beasts seem somehow to tessellate with this rugged northern landscape. They have big characters. Some are boisterous or flighty, others calm or retiring. It's fun getting to know them and their quirks, but I like them best for what they are doing to our land.

Cattle and ponies, with their larger mouths, aren't able to be as choosy as sheep are. They tend to rip vegetation, taking chunks out of whatever happens to be there, making it harder for coarse species to take over. Their greater bulk means that their hooves create gaps in the vegetation, opening the soil, enabling new growth. Their guts and their coats are a vector for seeds, helping plants to move around the landscape. Even their dung is important. As long as they are not pumped full of insecticides and wormers, the dung of livestock is a vital resource for a whole guild of specialist invertebrates, which are themselves the food of many species of birds and other creatures.

We're increasingly thinking of our livestock as conservationists, rather than just as meat on legs. Sheep have a role to play too, of course. They do a great job of grazing the regrowth in our meadows that comes up after the hay cut. Their ability to nibble grass right to the ground means that they strip more nutrients from sward and soil, which is exactly what many of the fragile meadow flowers need to prosper.

Putting the right animals in the right places at the right times is allowing our habitats to thrive. We're not just a sheep farm any more – we are edging back towards the more traditional mixed-farming model that was the norm until just a couple of generations ago.

In our enclosed land in Swindale and Naddle, fences and walls mean that we can apply different grazing regimes with ease. Ungrazed areas sit alongside meadows and pasture for sheep, cattle and ponies. This flexibility is what most farmers have within their gift.

On our big, open, windswept commons, though, managing grazing with the same degree of precision is not so simple. There are no physical boundaries between our mountaintops and bogs (which don't want grazing) and our grassy slopes and bracken beds (which do). These fells are about as close to wilderness as anywhere in the UK – but they're not wild, and simply walking away from them and letting nature take its course will have a mixed impact, enhancing the highest-altitude habitats, but likely allowing things to decline even further lower down.

All our native species evolved in a dynamic interconnected wilderness, where roving herds of herbivores followed seasonal abundances of food and were themselves followed and kept on the move by their predators. The unproductive uplands, with their high-altitude montane heaths and scrub, would have been largely left alone, as would the peat bogs, not being worth the effort of the climb or the risks of getting mired. Removing all human modifications and bringing back the full suite of native herbivores and predators to restore a true wilderness clearly isn't a realistic prospect in a place where we humans still make our home and our living, but perhaps there are ways to create a modern-day analogue on our fells.

If the right numbers of native herbivores are given the freedom to roam within a large enough area, nature will respond. The Lake District's ancient herbivore guild comprised cattle, ponies, deer and pigs. The deer are still here, and although the ancestral cattle and wild horses – the aurochs and the tarpan – are long since extinct, their domesticated descendants live on. Hardier, traditional breeds like Galloway cattle and fell ponies behave in a way that is as close to their wild ancestors as we can get.

Having seen how well nature had responded to the cattle in Naddle Valley and over at Ennerdale, and how rich the montane habitats had become in Carrifran and Norway, it was clear that grazing on our fells needed to change in order to make them truly rich again. However, because of the complexities of common land, making the change was far from easy. On Mardale Common, it took years of careful dealings with our neighbours, United Utilities and even the Secretary of State, but eventually our chance arrived.

Extending to 1,377 hectares, Mardale Common includes the most truly mountainous areas of our land. It occupies the high ground between Naddle and Swindale, and then flanking the lakeshore road to the south, it cups the southern tip of the reservoir. The two tarns Blea Water and Small Water fall within Mardale Common, as does Harter Fell, and its precious alpine flora.

There are only three farms with common grazing rights on Mardale Common, and Naddle and Swindale are two of them. Over the years I've established a good relationship with the only other commoner, who now shares our vision for its renewal. Most Lake District commons have many more commoners, so reaching consensus can be hard – this is a unique opportunity.

In the autumn of 2020, our 450 sheep were gathered from Mardale Common for the last time, for a while at least. Until the habitats on Mardale have fully recovered, and become rich enough to sustain a degree of sheep grazing again, the flock will only graze the enclosed land within our farm boundaries. Based on how long it took for the landscape in Norway to recover after the removal of sheep, I envisage this break lasting a few decades at least.

The year after the sheep were taken off, we rolled out the first iteration of a more natural grazing regime. On a warm day in August, with a small team of us gently driving them, eight belted Galloway cows and their calves walked up onto the common. I'd love to know what they were thinking when they set eyes on their home for the next few months. It would be the first time they had seen such a huge expanse of open land, probably the first time they had been at such an altitude. They probably weren't as interested in the views as about where the tastiest grazing was to be found. As cattle often do when introduced to a new piece of ground, they forgot about us, and went exploring. We wandered back down the hill, discussing who would take the first turn to check on them the following day. Four fell ponies, bought from one of the neighbouring farms, followed the cattle onto the common a few weeks later. Like the cattle, they quickly melted into the landscape.

Some might think that twenty or so animals is a ludicrously small number to be grazing such a large piece of land. Compared to a conventional farming operation, it is. But this is not a typical piece of farmland. There's a huge proportion of Mardale Common which I'm hoping won't be grazed at all. The mountaintops, bogs, bare rock, scree and the crags offer nothing in terms of grazing, and, given the choice, it's likely that neither the cattle nor the ponies will spend any time in them. These are exactly the places that will benefit the most from a total break in grazing.

With such a small herd, living out in tough terrain and rough weather, our cattle won't be very productive in a commercial sense, but they'll still yield produce and income. As they calve, the herd will slowly grow, but for a few years at least, low numbers are what we want. The ideal would be to vary their density from one year to the next, mimicking the natural cycles of boom and bust that would have affected their wild forebears. Using some new technology, we can simulate this too.

Our cattle wear GPS collars, the latest high-tech agricultural accoutrements. The collars communicate with an app, in which we can draw 'invisible fences'. As the animals approach the coordinates we've defined, their collars start to emit a beeping sound. As they draw closer, the beep gets more insistent. If they don't turn away, the collar gives them an electric shock, similar to that which they'd receive from a standard electric fence. I've been zapped by electric fences plenty of times. It isn't pleasant, and certainly painful enough that you don't want it to happen again, but it leaves no lasting impact. The shocks from the collars won't be nice for the cattle either, but it's more humane than the risk of a fatal tumble off a crag or a lingering drowning in a bog. Cattle are otherwise smart, and after a short period of training, ours all learned what the warning beep means, and they now almost always turn away from the invisible line well before the shock comes.

As well as keeping the cattle away from hazards, invisible fencing allows us to target their grazing. We can keep them away from areas that need more rest, like the sensitive mountaintops, or the places we've earmarked for montane scrub planting. Or we can concentrate them into bracken beds to be trampled, or places where dense mats of grass need opening up. The system also allows us to see where the animals are, in real time, making the job of checking them a whole lot easier.

What the invisible fencing doesn't do is stop sheep straying in from neighbouring commons. For that, we've had to rely on a more traditional approach: an actual fence.

It's possible to walk for many miles on the unenclosed commons of the Lake District without encountering a fence, or any other piece of human infrastructure, something which is hard to do in most other parts of the English countryside. In order to protect this cherished open character, putting up a fence on common land requires permission from the Secretary of State.

Accepting that we couldn't change how our neighbours farm on the common adjacent to us, we realized that the only way to achieve our ambitions for nature without impacting on them was through the installation of a fence, and so we launched into the tortuous application process. There were weeks of work, form-filling, landscape and heritage assessments, engagement with a list of prescribed consultees and notices put in the paper and at all the entrances to the common, to allow the public to have a say. Once it was all submitted, it took months before we got the go-ahead. Considering that the government claims to want to see more trees, wants more carbon to be locked up and wants to help wildlife – all things that the fence will enable us to achieve – they don't make it easy.

The new fence was installed in 2021, just before the cattle went out. It runs for a little over half a mile, from the upper edge of the Mardale exclosure, up and around Harter Fell, close to where the tea-leaved willows grow, to connect to a pre-existing fence that separates Mardale Common from neighbouring Longsleddale. The remaining eastern part of the common is already fenced against the upper farm boundaries of Naddle and Swindale. By connecting into these other fences, we're creating what is in effect a 930-hectare, high-altitude field, in which the cattle and ponies roam, but where the neighbouring sheep are kept out.

We had to compromise. Some of the consultees involved in the application process made it clear that our proposal to fence the entire common was unacceptable, believing it would have a negative impact on the visual appeal of the landscape. So the fence we were given permission for only protects two-thirds of Mardale Common. Neighbouring sheep will still be able to stray onto the remaining 400 hectares at the Common's western end, but we've been living with that for the last decade, so while it's disappointing not to be able to help this area recover too, it's no new hardship.

Fences are an obvious way to enable huge improvements to habitats across the Lake District fells. It's true that they can have a slight impact on the visual character of the area, but this can be dealt with through sensitive landscape design, hiding the fence in the terrain. And yes, they can have a slight impact on access, but this can be dealt with by installing gates and stiles, as we've done in several places in ours. Our new fence will give wildlife a massive boost and avoid issues for our neighbours. Minor access and aesthetic impacts feel to me like prices worth paying.

I love the fells as much as the next person, but I'd love them even more if they had more wildlife. A few hundred metres of posts and wire will allow us to do something very different to our neighbours in a way that respects their desire to continue to farm as they have been. This is a situation where good fences really will make good neighbours.

There are critics who say that by no longer grazing sheep on Mardale Common, we've failed in our attempt to balance conservation with hill farming. I disagree. We still have livestock on Mardale. Hardy cattle and fell ponies are every bit as important in the traditions of Lake District farming as sheep. I don't expect everyone will follow our lead, and I know that the changes we're making aren't universally popular. The fences upset some people. Others used World Heritage Status to argue that the fells should always have sheep on them.

For every vocal critic of the changes we've made, I suspect there are a hundred silent supporters and a thousand people who couldn't care either way; not everyone thinks about nature and the landscape so deeply. It's an irritating quirk of human nature that we seem to be far more ready to complain about the things we don't like than we are to extol the virtues of the things we do.

Once again, I was in the firing line, but it felt noticeably different this time. Perhaps it was because the criticism was for things that we were actually doing, rather than for things that people thought we were going to do. Perhaps it was because we'd explained why we were doing it more effectively. Perhaps it was that people now knew who we were, who I was, and that the decisions we were making were ones with people behind them, rather than a faceless organization. Or perhaps it was because we weren't so alone in making these sorts of changes – there are farmers and organizations managing their land for nature all over the Lake District now. Whatever the reason, the flak that I was braced for didn't really materialize. There were a few sniffy comments accusing me of imposing an 'extremist rewilding agenda', but this time around, safe in the knowledge that I was just one of a growing community doing similar work, they didn't get me down.

So, with the sheep gone, what will happen to Mardale Common? In short, it'll get scruffier, scrubbier, wetter and wilder. Mardale Common has variable terrain, running from around 250 metres above sea level where it follows the road along Haweswater's eastern shore, up to over 800 metres at Racecourse Hill, two miles away on the common's western edge. Within this range lies a complex mosaic of valley mires and rocky knolls, patches of ageing juniper scrub, steep crags, tarns, gills, heath, grass and bracken.

We have already largely sorted our peat bogs by blocking up all the drains that had been cut into them over decades past. They should now be able to look after themselves, staying open, since they're too wet for trees to grow. The carpets of sphagnum mosses and other specialist plants that can tolerate the waterlogged conditions will build over time, gradually laying down new peat as they grow, locking up carbon and providing a home for wetland life. On drier ground, particularly in areas dominated by bracken, we are planting birch, rowan, holly, hazel and oak. Higher up, juniper and downy willow are going in, to restore the long-absent montane scrub. The dispersed grazing of the cattle and ponies will mean that the flowers will move back in, and the other wildlife will follow.

Hill farming and commoning has always been about providing for society's needs. As these needs change, so will the farming. Our management of Mardale Common will be different to what's come before, but that can be said about almost any point in Haweswater's history. We will still be a part of the Lake District's cultural landscape. We'll still have distinctive native breeds of livestock, tended by local people, delivering benefits to the community and the economy, just as Lake District farming has always done. I'm comfortable that what we are doing fits perfectly within this ever-evolving story. This is simply another step along that endless road. I can't wait to see how nature responds.

CHAPTER 16

Fields

NADDLE:
The wedge-shaped valley
(Old Norse)

The track leading up to Naddle Farm is flanked on either side by our holding fields. Being so close to the farmyard and its pens, these fields are where the sheep are kept while waiting their turn for shearing, medicating or sorting. Most farms have a few fields like these, where it's difficult to make any real conservation gains. Our holding fields get heavily grazed for short periods during the spring and summer. This prevents most plants from flowering, so like a regularly mown lawn, they aren't interesting botanically.

The close cropping has another, unseen impact. Where heavy grazing keeps grass short, the underlying soil is often seriously compacted. When I camped with Becki and the kids in the middle of one of the holding fields a few years ago, I bent almost every tent peg – the ground was more like concrete than soil. Soil compaction is partly due to the physical trampling of the livestock, but mostly it's down to a lack of root depth. A tall plant needs deep and spreading roots for stability, as well as to harvest the water and nutrients it requires to sustain its bulk. Conversely, plants kept short by grazing tend to have shallow roots.

Our holding fields aren't totally without wildlife interest, though. Rising out of them, like statues of some goddess of nature, are a few dozen craggy oak and ash trees, their trunks thick with life and age. Trees grown in open conditions like this get the chance to spread their mossy limbs wider than in woodland. They become the best a tree can be, giving shade and shelter to the animals in the field below and a home to redstarts, purple hairstreak butterflies, epiphytic ferns and lichens. Their roots penetrate to great depths, drawing on subterranean stores of water and nutrients that other shallower rooted plants can't access. These, along with carbon-rich sugars that the trees have manufactured themselves through photosynthesis, are available to the animals grazing and browsing on them. In the autumn, the trees' falling leaves bestow a huge dose of goodness to the topsoil and a feast for fungi, worms and other decomposers.

As timeless as they seem, no tree lives for ever, and several

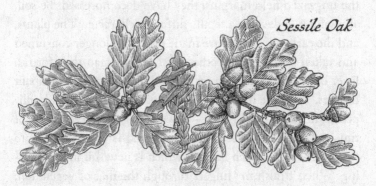

Sessile Oak

have fallen into the fields along the track in the years I've been driving up and down it. Dotted around the holding fields are a scattering of 2 x 2 metre wooden cages, into which we've planted the next generation of trees, standing ready to take over from their elders when their time comes. Inside the tree cages, and in the strips at the field edges where we've fenced the animals out to establish new hedges, luxuriant plant life now grows, in sharp contrast with the short green turf of the adjacent grazed areas. Tall wildflowers and grasses sway against the stems of the planted trees. This lush above-ground growth is mirrored below by deep penetrating roots.

We don't generally get to see roots, or to witness their incredible strength, but there are places where their power can be appreciated. Where trees grow on crags or cliffs, their roots can sometimes be seen worming their way into fissures, in search of both water and purchase. Over time, as these roots grow, they can force these fissures apart, and in some instances split great boulders in half, or cause sections of crag to fall. Although rarely visible, this enormous cleaving force is constantly at work beneath our feet.

As roots have exerted their power below the tree cages and the ungrazed field margins, they have decompressed the soil, allowing it to develop a totally different character. The plants, and the carbon that they are made of, are no longer consumed and taken away in the bodies of sheep. Instead they form a layer of plant litter, made up of dead leaves, stems, and spent flowers. With the help of worms, woodlice, springtails, millipedes and other detritivores – as we biologists call this team of tiny recyclers – this dead plant material is broken down to form new, carbon-rich humus, which is new soil in the making. When I push my fingers through the mat of vegetation, the earth below is rich and crumbly, dark and fragrant.

Outside of the cages, in the body of the field, getting into the earth without a spade or fork is impossible.

The porous structure of these de-compacted soils can hold more water. If soils were restored everywhere, they would provide massive flood relief, like spreading a huge sponge across the landscape. With tiny pockets of air and water within their loamy matrix, healthy soils allow a huge range of fungi, invertebrates and microbes to flourish. This unseen web of life is constantly cycling nutrients and making them available to the plants, boosting their nutritional value to the herbivores.

The greatest potential benefit of healthy soils, however, is to the climate. There is more carbon in the soil than there is in the atmosphere and all the world's plants and animals combined. Our management of the land can either build up or degrade this mighty carbon store. Where land is sensitively managed and the web of life in the soil is properly functioning, decomposing plant tissues are converted into new soil, locking the carbon they are made of into the ground. Where soils are degraded, compacted, overgrazed and overworked, they switch from being a carbon sink to a carbon source. Trees get all the press as a tool to fight climate change, but soils have the potential to be even more valuable if we care for them properly.

Agriculture contributes around 10 per cent of the UK's greenhouse gas emissions. Although this figure is dwarfed by the amount that comes from transport and electricity generation, every sector needs to play its part. The National Farmers Union believes that UK farming can become carbon neutral by 2040 and is supporting farmers in achieving that goal. Most of the greenhouse gas emissions from farming are in the form of methane, produced by livestock as a by-product of digestion, but fuel and fertilizer use also play a big part. Achieving net zero in farming will require reducing emissions, including livestock numbers, while simultaneously capturing more carbon. Improving soil health will be a big part of that.

Thankfully, the importance of caring for soil is rapidly rising up the agenda and is one of the areas where farmers and conservationists are increasingly aligned. Regenerative agriculture, which has soil health as its primary goal, is gaining traction among farmers. It is to farming what rewilding is to conservation – innovative and exciting – and the two approaches have much in common. The concept of rest is at the core of regenerative agriculture; a break from grazing can rapidly repair damaged soils and deliver many benefits, both to the environment and to farm businesses. In some situations it seems that pulses of grazing, interspersed with periods of rest, are even better for soils than rest alone. Grazing is a natural process after all, and the soil, microbes, fungi, insects, plants and animals are all part of the same living system.

There's no one right way to go about getting the benefits that healthy soils provide, and at Haweswater we're using a range of methods. In some places, especially where the ground is particularly steep, along watercourses or around our existing woodlands, we're excluding grazing altogether, giving the soils extended periods of rest. On the higher ground, on the commons, we're giving the cattle and ponies a degree of freedom to decide which areas they think should be rested. By keeping their numbers low, some areas will be grazed and others ignored, allowing the land to develop a mosaic of habitats. On the land within our farm boundaries, we're following the lead of some of our farming neighbours, and taking a more active approach.

Sam and Claire Beaumont, who farm a beautiful wedge of land rising from Ullswater's shore, are employing a technique evocatively known as 'mob grazing'. Using electric fences, they concentrate their herd of shorthorn cattle into narrow strips of pasture for short periods before moving them onto the next strip, allowing the previous one to rest. This regime simulates large roving herds of herbivores moving across the land, grazing intensively, then moving on. I attended a regenerative farming course at Claire and Sam's farm a few years back, and as I pass their place on the school run, I pop in from time to time to pick their brains.

Regenerative farming and mob grazing are systems that have much in common with traditional farming practices. An old farming saying goes 'never leave the sheep in the same field long enough to hear the church bell ring twice', meaning that sheep shouldn't graze in the same place for more than a week. The patchwork landscape of numerous small fields, which existed before thousands of miles of hedgerow were ripped out during the years of post-war agricultural intensification, was one well suited to this ethos.

Paul and Nic Renison are restoring this patchwork at Cannerheugh, their farm on the edge of the Pennines, twenty miles north-east across the Eden Valley from Haweswater. I've only ever visited them on still days, but in the winter Cannerheugh is often pummelled by the Helm Wind, a locally infamous and fearsome north-easterly, and the UK's only wind to have a name. Nic and Paul, working with Pete from the Woodland Trust – he really does get around – have planted hundreds of metres of new hedgerow across their land. The shelter they provide not only gives a windbreak to their cattle, sheep, pigs and chickens, but it helps the grass to grow too – exposure to the wind seriously hampers plant growth. By dividing their fields with new hedges, mob grazing becomes easier. The benefits to nature that the hedges provide are an accidental upside.

As their soils recover, Claire, Sam, Paul and Nic are seeing impressive increases in the growth of grass without the need for synthetic fertilizers. The periods of rest between the bursts of grazing allow the grasses to grow taller, and to put down deeper roots. Deeper roots mean healthier soils, and healthier soils mean better plant growth and more grass. If the periods of rest are at the right time, the approach also promotes greater diversity, allowing plants to flower and set seed. Greater diversity in the pasture equates to a more varied diet for the livestock, keeping them healthier. This natural feedback loop translates into better-quality farm produce, reduced costs and more profit. Better for wildlife, better for the ecosystem, better for livestock and more productive – a focus on soil is a no-brainer. We're a few years behind the curve at Haweswater, but now that we're only keeping our sheep on the enclosed land within our farm boundaries, regenerative farming is where we're heading. Nic, Paul, Claire and Sam are among a growing number of regenerative farmers that we can take inspiration and advice from. It's an exciting community to be a part of.

Accounts of the many benefits of regenerative grazing practices are compelling but, so far, many in academia and government are retaining a degree of scepticism. I guess that's understandable. Changes to soil carbon in response to management can be tricky and time consuming to quantify. This is one of the things that Ash, our conservation scientist based at Haweswater, is trying to remedy. The long-term grazing experiment that she is establishing will contribute to plugging the evidence gap.

To my simple mind, any system that uses fewer artificial

fertilizers and allows the land to rest has got to be better than the more intensive systems that have dominated over the past few decades. At Haweswater, there's no doubt that the improvement in our soils, the tree planting, the bog restoration, and the shift to a low-input/low-output livestock system, are working together to deliver a whole range of benefits to nature, water and the climate. I hope the science will catch up soon, and be enough to convince government that regenerative farming is the best type of farming to support.

It turns out that farming with nature is as good for farm finances as it is for the planet. The Lake District farmers that I know are not farming just for the money. For many, food production is a cause, a way of life, just as caring for our land to enhance its value to nature is for me. Farms still have to function in the real world, though, and without reliable income from somewhere they wither and die, just like any other business.

When we started at Haweswater, we committed to running the farm with an open-book approach to our finances. We published the economic report for our first three years in 2016 when we had over 1,500 breeding ewes. It made for sobering reading. It showed that on average farming livestock made a loss of £162,131 per year, before government grants were factored in. The £55,272 of income generated from the sale of our livestock was dwarfed by the £217,403 cost of producing it. It is only the annual investment of almost £300,000 of government grants that made the operation stand up. The money left over after paying for the farming enterprise covers the salaries of my colleagues and me, the rent we pay to United Utilities and the conservation work that we do. In most years, we operate in a way that is cost neutral to the RSPB, much as an independent farm business would, spending the income that we generate in the local area.

In 2019 a ground-breaking report entitled *Less is More* was published. Commissioned by the RSPB, National Trust and the Wildlife Trusts, the report presented an analysis of 46 upland farm businesses, one of which was ours at Haweswater. It highlighted the vulnerability of hill farms and their extreme dependence on public payments to make their businesses sustainable. What was made very clear was that our finances were not unique. In upland areas like the Lake District, where high rainfall, low temperatures and low plant productivity make any form of agriculture a challenge, the only way to stand a chance of making a profit is to farm less intensively, and crucially, without artificial inputs.

At first glance, this runs counter to logic. You'd be forgiven for thinking that more livestock would mean more profit. In fact, past a certain point, individual to each farm, more livestock often means more loss. As soon as a farmer tries to compensate for the natural deficiencies of weather and altitude and exceeds the number of livestock that the natural vegetation can sustain, they will start to lose money. It becomes a vicious cycle: the more money that is spent on fertilizer, supplementary feed, expensive machinery, buildings for winter housing and so on, the more will be lost. In essence, the report recommends farming in a traditional way, relying purely on what the earth and the sun can freely provide. This is what is meant by a closed farming system, one which relies only on the farm's own produce, and not on inputs brought in from elsewhere. Until a century ago, this is how all farms had to operate.

This principle now guides our approach to farming at Hawes-water and explains why we've reduced the overall size of our flock. The aim is to make the farm fit the landscape again.

Having fewer sheep has drastically reduced our costs as well as our carbon footprint. Having all our sheep contained within the farm boundaries means we can keep a closer eye on them and ensure higher standards of welfare. The public are becoming ever more discerning about food, where it comes from and the impact that it has on the environment. I'm hopeful that the role our livestock play in the regeneration of nature will make a compelling reason to buy beef and lamb from us.

Of course, changing a farm business can't happen overnight. Everything in our operation is interconnected. We couldn't reduce the size of our flock until our agri-environment agreements allowed it, and so for a while we were stuck with a minimum flock size that was still too big. We needed new fences to enable our new grazing plans to work without affecting others. Each step takes time, but we're moving in the right direction.

Tradition is important in the Lake District, but it's a slippery notion, and there's no one time that everyone can agree on as being the moment when farming was perfect. Probably it never was. That said, there is a massive amount to be learned from the past. Much of what we've done, and what we plan to do in future, is informed by history. The ancient principle of 'levancy and couchancy', meaning 'to get up and lie down', was used by landlords to ensure that farming tenants only carried the number of livestock that the landscape could sustain – basically the same principle as lies behind the *Less is More* report.

We make one compromise to this closed-farming system, though it's one that also has a historical precedent: we send a proportion of our sheep away to lower ground for the winter. Before the widespread availability of synthetic fertilizers, lowland farms would manage their soil fertility with more benign methods, using different crop rotations, rather than growing the same crops year after year, as is common practice now. These rotations often incorporated grazing, which injected a dose of fertility into the soil via the livestock's dung, rejuvenating it in readiness for the next arable crop. Our sheep take their winter holidays on another RSPB nature reserve, where they create sward conditions perfect for breeding lapwings and other wading birds.

Many other aspects of our day-to-day farming activities are done the traditional way. There are no technological fixes for maintaining the many miles of drystone wall on our farms, and Spike spends a massive amount of his time labouring to keep them upright. We still use many of our stone barns to store hay in, or to provide animals with winter shelter. We manage our hay meadows the way they used to be managed: with a single late-summer cut and without fertilizer. The breeds of sheep, ponies and cattle on our farms are those that have been part of the landscape for centuries. In many respects, we are farming in a way that was widely practised up until the 1950s, when farming and nature got along more comfortably.

There are some elements of this historic system that we can't recreate. The slow and steady selling of land and farmhouses has seen the number of farms and flocks reduce, with those that remain having grown by subsuming the surrounding land and livestock. Large modern farm buildings clustered around most traditional Lakeland farm steadings tell this story with concrete clarity. Before this change in the pattern of farms and prior to mechanization, farms casually or permanently employed many more people to carry out the back-breaking toil involved in keeping the countryside productive: cutting meadows with scythes; turning and taking the hay to the barns; ploughing with horses or oxen; planting and harvesting cereal and root crops; cutting bracken; burning charcoal; harvesting peat; close shepherding the livestock on the fell through the summer months. All these tasks were carried out by armies of local people. That everyone was involved, even the children, is shown by the fact that our long school summer holidays originated in an age when children needed the time off to help with the hay harvest. Now, of course, one tractor can do the work that ten people used to do, and in a fraction of the time.

I find the idea of de-mechanizing the farm – replacing tractors and quad bikes with scythes, ponies, backs, hands and feet – a tempting one. It would drive a reconnection with the land for all the people that would need to be involved, as well as reducing our impact upon it. But it isn't realistic. There isn't the money in farming to pay the wages that such a huge workforce would require. Getting twenty or thirty people to drop everything to come and help cut our meadows when the weather comes right isn't going to happen.

That said, there are a lot more people involved at Haweswater now than there were in recent decades. There are the same number of farming staff as before we took over, but we've added an eight-strong team of full-time RSPB staff, myself included. We have both local and residential volunteers mucking in as well as academics carrying out research. We use local contractors for habitat restoration, tree planting, fencing, building and repair work. Naddle and Swindale are thriving, busy places these days. There's often concern among rural communities that a move away from familiar farming practices will lead to depopulation, to the closure of schools and pubs and the abandonment of land. Our set-up shows that this doesn't need to be the case. My colleagues and I are as much a part of the rural community as anybody else.

There's a lot of work involved in getting nature back onto a strong footing, and plenty of spending. We've brought in almost a million pounds from various charitable sources to carry out habitat restoration work over the past decade. That's in addition to the three million pounds that we've received in government farming support and agri-environment grants over the same period. That's a lot of money invested in employing people, restoring the land and supporting the local economy.

As natural processes reassert themselves, we'll be looking to nature to do more and more of the heavy lifting as time goes on, but there will always be lots for people to do at Haweswater. Where we don't get the natural regeneration that we're hoping for, we'll continue to plant trees. Hedges will still need laying, walls and fences will need repairing. The livestock will need looking after. Deer populations will need to be managed.

Bringing back some of the Lake District's shadow species will take work too. Feasibility studies are being carried out to identify what needs to be done to return pine martens, beavers, corncrake, black grouse and others to their former haunts. Pine martens will help keep invasive grey squirrels in check, providing relief for the native reds. Beavers would add hugely to the work we've already done to slow the flow, creating complex, water-filtering wetlands that teem with life. Black grouse and corncrake will be barometers, their presence showing that the habitat mosaics they rely on are in good shape.

Ten years ago, reintroducing missing species to the Lake

District felt like a remote prospect, likely to meet insurmountable opposition. Things have changed. There is growing support for the recovery of nature in all its forms everywhere I look, including from farmers. Species reintroduction is explicitly mentioned in the Lake District National Park Plan, something I never would have expected a few years ago. The real challenge now is to harness the growing enthusiasm and translate it into action, slotting these missing pieces of the ecosystem back into place.

As nature reasserts itself, there will be more colour, more movement, more of the sounds and smells of the living world for visitors to enjoy. There aren't many places in the Lake District offering opportunities to connect with wildlife, so we're branching out into ecotourism. We started small with guided walks, taking groups into the woods to listen to the dawn chorus, into the meadows to revel in the flowers, or to watch red deer stags battling for dominance, roaring into the echoing fells. It was clear that there was demand so, in 2019, Heather, naturalist and entrepreneur, joined our little team to develop more ideas for what we could offer to visitors. Heather has an instinct for knowing how to share nature with people. As well as working at Haweswater, she runs a social enterprise with her partner Cain, running family-friendly activities that combine nature with nourishment. Moths & Muffins, Frogs & Flapjacks, Bats & Pizzas, and Kittiwakes & Doughnuts events have helped thousands of people to appreciate the overlooked beauty of the wildlife on their doorsteps.

Heather's infectious enthusiasm has got Haweswater's ecotourism off to a flying start. Shortly after she started, we opened a badger hide at Naddle Farm. With some careful peanut placement, the local badgers soon learned that it was worth visiting as part of their nightly foraging rounds. It's been a huge success, with the slots to book it always selling out. Some 250 people booked in the first year, and the costs to build the hide were recouped in half that time.

Heather usually does the guiding in the hide, her lilting north-east accent quietly telling the visitors all about our badgers' secret lives, while waiting for the sun to sink and the stars of the show to arrive. I've done a few sessions and get as much pleasure from seeing the badgers as the visitors do. On my first one, after a tense half hour wait, the cartoon faces of two badgers burst into the illuminated viewing area in front of the hide. For the next forty-five minutes, as they rootled around for peanuts hidden under rocks and logs, at times less than a metre from the glass, the six visitors and I were entranced. A smartly dressed woman in her forties couldn't stop saying 'wow' over and over, and another was close to tears. None had ever seen a badger alive before, let alone at such close quarters. Watching a big, enigmatic creature like a wild badger can be a powerful and moving experience. It's a privilege to be able to provide these opportunities for people, helping them to build an emotional connection to nature. The fact that our hide also generates income that we can invest in sustaining the landscape is the icing on the cake.

We're building more hides and have plans for a small bunk-house in one of the old barns. It will all be low key, in keeping with Haweswater's tranquil, backwater character, and there will be plenty of places where the creatures are left to their own devices, unlooked at and undisturbed.

By establishing a successful, diverse business, with income from farming, grants and tourism, we'll be more resilient to whatever curveballs the future may throw. Livestock is only one of our products now. With the concept of public money for public goods at the heart of the government's new agricultural support system, it is farming for water, wildlife, carbon and beauty, as well as meat, that will sustain us. Our customers are no longer just the people who buy our sheep and cattle, they are the visitors enjoying a nature-rich walk through Swindale's meadows, or enraptured by their first encounter with a badger. Our customers are the communities at risk of flooding downstream, even if they don't know it. They are the people drinking tap water that once fell as rain on our hills, and the people breathing air with slightly less carbon dioxide in it. You are one of our customers. These benefits flow to you as a result of the way that we care for our land. For those goods for which there is no accepted cash value, government pays us on your behalf through grants, and my job is all about ensuring that you get a good return on your investment.

To run a diverse enterprise like this relies on a diverse, healthy landscape. This means embracing a whole spectrum of different land-management approaches. For a while, rewilding contributed to a widening of the gulf between farmers and conservationists, a false dichotomy suggesting that you could commit to rewilding or to farming, but not both. The truth is far easier to stomach than that. It doesn't have to be either/or. There's plenty of room in our islands to increase the land given wholly to nature, to expand sustainable, nature-friendly farming, and to do a whole range of things that sit somewhere in between.

Diversity at the National Park level should be easy to achieve. Every farm in the Lake District is distinct, each with its own set of physical, climatic and environmental constraints and opportunities. We farmers and land managers are as varied as the landscape, each with our own passions and priorities. Keeping up with the ever-changing demands of society to find a way to care for land that is right for the time, the place and the individual means that there are already a vast array of land-management approaches in the Lake District, each with their own merits and pitfalls.

Much of what we are doing at Haweswater could be applied more widely, and I'd be lying if I didn't say that I hope that more farmers might make similar nature-focused decisions, but that's for them to decide, not me. Our way is not the only way to care for land, but we are doing what we believe is right and others can watch as we stand or fall.

First and foremost, I need to do the right thing here, where I'm standing. Leaving Haweswater in a better state than when I found it is all the legacy I can really ask for.

CHAPTER 17

Fragility

BRANT STREET:
The steep path
(Old Norse/Old English)

We are sometimes accused of going too quickly at Haweswater, of pushing for change at a pace that might scare or alienate our neighbours. But with wildlife and the climate collapsing so rapidly, time is a luxury we don't have. We have an opportunity here, one that needs grasping with both hands to stop it slipping away.

Yet like all other farmers, we're at the mercy of both the state and our landlords. Although government is making positive noises about supporting nature-friendly farming, the details are painfully slow in coming, and there's little to stop the emphasis shifting again with the next election cycle. And I try not to think about what might happen if our landlord has a change of corporate policy.

Meanwhile, strides are being made in the manufacture of synthetic alternatives to meat. On climate grounds, this may prove to be a game changer, but it might also kill off livestock farming, along with livelihoods and the potential to use sympathetic farming as a force for ecological recovery. Some might think that is a price worth paying, and that we'd all be better off as vegetarians. I'm not so sure. We should certainly not be eating meat if it leads to rainforest clearance, run-off into rivers, or animal mistreatment. Buying less, but better meat, in order to encourage low-impact, wildlife- and climate-enhancing farming feels like it could be an important part of the solution.

But, unless we get our collective act together quickly, there's a very real chance that the changing climate could undo all our best-laid plans. Our alpine plants might be out-competed by their more vigorous lowland competitors, as the conditions in the crags warm up. Wetter summers might prevent us from cutting hay. Winter storms might destroy our fences and walls and make our valley bottoms unmanageable. Our bogs might dry out and catch fire in heatwaves. Stressed by the climate, more of our trees might suffer from fungal diseases, as ash and juniper are already experiencing. We need to make our land-scape resilient to all these threats as soon as we can, before it becomes impossible.

Juniper

And so, informed by the past, but with eyes to the future, renewal at Haweswater is underway. It's a process of constant learning. Nobody has all the answers. I lie awake at night weighing up our options, hoping that our solutions are the right ones. We've achieved a lot, but there's still a long road to travel. As the problems continue to evolve, so will our responses. Some battles will be won, others will not. There will be joy and sorrow in the trying.

I've developed a deep connection to this place, as have the many friends and allies I work alongside. It took me a long time to realize how important local relationships are to our endeavours, how essential respect and dialogue are to achieving anything. I don't expect to get along with everyone, but I've learned the value of seeing the whites of people's eyes, of hearing other sides of the story. The lessons I've learned have shaped our plans and helped me understand how we can fit into the sprawling, ever-changing epic drama of the Lake District. Like the tree planters at Carrifran, we are all playing our part in a collective effort, investing ourselves in the land, contributing to a shared legacy.

Haweswater is unique. But then so is everywhere. Everything we've done is specific to our context. On the region's most important supply of drinking water, and with nature conservation in our blood, our motivations are different to those of our neighbours. The creation of the reservoir decades ago changed the land for ever, submerging farms and homes, sweeping aside traditions that were practised by the people who came before us.

I like to think we're creating new traditions. Others with different ideas may come after us, but the marks we're making are indelible growth rings in the heart of a tree. E. O. Wilson, the legendary biologist, said that 'there can be no purpose more enspiriting than to begin the age of restoration, reweaving the wondrous diversity of life that still surrounds us.' Perhaps this new age will show up in peat cores taken by future historians, a period when flower and tree pollen begin to rise again.

Wildflowers are our weathervane. As they proliferate, they become symbols of change, of hope, of further richness to come. With each passing year, there are more flowers at Haweswater. The insects, the birds and the rest of the wild horde will follow.

CHAPTER 18

Future

WOOF CRAG:
The rocky heights frequented by wolves
(Old English)

It is 2050. Decades have passed and after half a lifetime at Haweswater, I'm a few weeks from retirement. It's a perfect late May morning, and I'm walking from Naddle Farm up to the top of Selside Pike, one of the summits on Mardale Common, perhaps for the last time in my working life.

Lambing finished a few weeks ago, and most of our flock of 200 ewes are in the fields around the farmyard with their lambs. The first spring flowers are showing and there's a good growth of grass, thanks to the months of rest that these lambing fields have had over the winter months. The sheep are taking advantage of the shade beneath a large spreading oak, one of many that stud these fields. A pied flycatcher darts into a hole in one of their broad trunks, a beak full of grubs for hungry chicks.

Jen, our young assistant shepherd, has finished doing the rounds of the sheep and is heading up the valley to check on the cattle. We walk together for a while; she doesn't seem to mind slowing to my old man's pace. I reminisce about the old days, when this farm carried eight times as many sheep as it does today. What a back-breaking grind it was, with the extra feed we had to buy in, the vet bills, the haulage, the unnecessary deaths. Jen raises an eyebrow, sceptical that things can ever have been so intensive. She's only in her mid-twenties, so her baseline starts about five years ago, after almost everyone had adapted to the low-input/low-output system that dominates now. Society's changing attitudes have helped farms like ours to market their produce; consumers not only expect meat to be produced to the highest possible welfare and environmental standards, but they expect it to be as local as possible. Fortunately, because they eat it less often, they are prepared to pay a good price.

A sleep-ruffled couple emerge from one of the treehouses we rent out. Half-joking, they complain about the rasping corncrakes that woke them at an ungodly hour and thank us for the complimentary earplugs. Yesterday, they encountered red squirrels, badgers, an osprey and a pine marten. The lamb and venison they ate at the local pub came from Naddle Farm, so we chat about the rotational grazing system that we use, and how a diverse diet of wild plants makes the meat so tasty.

Later, Jen and I enter the wilder ground of Naddle Valley, a natural bowl surrounded by woodland with a network of beaver ponds at its centre. As always, it's noisy with birds and insects. A red-backed shrike dives past and takes a dragonfly out of the air. We startle a water vole, which plops into watery cover. A red kite drifts across, but with more important things to think about, she pays us no attention.

Jen spots the cattle grazing on the far side of the valley, taking advantage of the tree cover, so she heads off to check on them. Not that they really need it. We accepted a while back that restoring species like bison to a place as full of visitors as the Lake District wasn't sensible, so we make do with belted Galloways. Even though they are very much farm animals, when they appear through the scrub in an unruly landscape like this, it feels like an encounter with something far wilder. They largely look after themselves, much as their wild forebears did.

This middle part of the valley is the bridge between the farmed land below and the even wilder land above. The sheep and cattle only come up here for brief pulses of grazing, varying their timing from one year to the next. It's our attempt at simulating the impact of the roving herbivore herds that our wild plants evolved alongside. It might look random, but we're careful to ensure that every patch of land gets the opportunity to have a spring and summer off at least one year in three, so that the wildflowers can bloom and set seed. This maintains the landscape's dynamism; there are always areas of short grass for black grouse to lek in, longer areas for barn owls to hunt, plenty of seeds, berries and nuts for finches, thrushes and squirrels to feed on, but these features are rarely in the same place twice.

A pair of teal splash out of the highest of the beaver pools, leaving a brief rainbow in their wake as I approach the gate that will take me up onto Mardale Common. When I started at Haweswater, the boundary wall that encircles Naddle Farm marked a stark transition, visible from miles away, with woodland below and open fell above. It's no such Rubicon now. For the last twenty years, trees and scrub have been reclaiming more and more ground. Jays must have been burying acorns up here for centuries, snaffled from the woodland below. It was amazing how fast the oaks appeared once the grazing was sorted. The common rights that come with our farm tenancies include pannage, the right to turn out pigs to forage for acorns. For the first time in centuries, this might soon be a right worth exercising again.

Birch and rowan trees moved in fast too, especially in the areas with heather, which provided perfect natural protection from grazing while they were vulnerable saplings. Most of the drier rocky knolls that rise up out of the peat bogs have a pretty decent covering of trees now, with a lush carpet of bilberry, cowberry and crowberry growing below. When I come up here with my grandchildren for late-summer bilberry picking, they think I'm joking when I tell them how bare it used to be.

One of my favourite places on Mardale Common is Woof Crag, a squat square block, floating like some granite warship on a sea of moss. It's not a celebrated landmark, just a 100-metre length of exposed rock, a small cliff to one side and a grassy slope that invites ascent on the other. It's always drawn me. This was the first place in England that bluethroats raised young for maybe a thousand years. Not so long ago, no one thought of bluethroats as a breeding bird in England. Robin-sized members of that family, bluethroats aren't much to look at from behind. But catch them from the front, and their blue, black and chestnut throats are unmistakable, as is their rich and complex song. They used to breed sporadically in Scotland, and their numbers increased as the Scots – always ahead of the English – restored their montane scrub.

As soon as our own montane scrub started to mature, the bluethroats arrived. The downy willows that I planted on the shoulder of Woof Crag one wet winter twenty-five years ago flourished, and now the bluethroats are drawn to the place in the same way that I've always been. There is a pair here again this year, and I let the male's virtuoso performance entertain me while I have some coffee from a flask. That the tiny willow

Downy Willow

cuttings I planted have turned into a tangle of habitat supporting a bird as beautiful as any jewel binds me to this place.

Hundreds of other people have their own connections to
this landscape. Although nature did most of the legwork, we
gave her a hand. The seed source was virtually extinguished
for some species, particularly for aspen and willow, but they
are properly established now, marking out the damp edges of
the mires. I long ago lost count of how many trees we planted,
and of how many volunteers helped with the planting.

Moments later, I bump into Spike near one of the aspens he
remembers planting twenty winters earlier. He's been retired
for ten years, but, having spent almost his entire working life at
Haweswater, he can't keep away. When we reintroduced sheep
grazing on Mardale Common a few years ago, Spike was worried. He was convinced that all of our efforts would be undone,
that the habitat would revert back to the grassy desert that it
was a quarter of a century ago. It hasn't. Following two decades
of sheep-free recovery, the alpine-style grazing we now use
enhances the habitats. Small flocks of sheep and the scattering
of cattle and fell ponies, grazing for short pulses, keep the coarse

vegetation from dominating and create niches for seeds to find soil, allowing new plants to grow. Just like in the Alps, the sheep are shepherded. Jen moves up into the hut we restored at Low Loup for two or three months in the summer, moving the animals around, keeping them out of areas that are still regenerating, and helping them to find the good grazing.

I've done a season or two up here myself. People pay to experience it, so there's often company. Outside the grazing season, we rent the hut to holidaymakers, or for longer stints to artists and writers looking for tranquillity. Low Loup looks out over Haweswater reservoir, still the most important source of drinking water in the region, but now also the water company's cheapest. The water flowing into it is crystal clear; the soil that used to be eroded into it year after year now stays on the hill. Diverse and luxuriant plant growth above ground is mirrored by a deep and stable underground network of roots, binding the earth in place.

Sitting beside the Old Corpse Road on a herb-cushioned boulder, I release the intoxicating scent of wild thyme as I break out some sandwiches. Bees and butterflies are on the wing. Green hairstreaks are busy in the bilberry and the ubiquitous orange tips are making the most of the last of the cuckooflower. A violet oil beetle lumbers out across the path in front of me, quickly followed by a common lizard missing its tail. I hear a ring ouzel in the scrub behind me, another species that appreciates the changes up here; their numbers have tripled in the last two decades.

A couple of walkers pass by and nod amiably. It always amuses me how faithful people are to known walking routes, but the constant foot traffic has meant that the path is kept open, so I'm not complaining. Although the landscape now is more tangled and complex in response to the move away from constant sheep grazing, it hasn't turned into impenetrable scrub as the nay-sayers said it would, and the popular footpaths are as accessible as ever.

I push on up the hill. The climb up to Selside Pike used to be such a thankless slog. For the first ten years that I surveyed this section of the fell, I would record nothing but meadow pipits and skylarks, and even then not many. The first changes in the birdlife came after the bogs were restored. Snipe returned immediately, drumming and chipping and then flying out of the moss like enraged bats. Red grouse arrived as soon as the heather got going.

The woodland and scrub birds came in with the habitat too. Cuckoos are a common sight and sound up here in spring, along with tree pipits, linnets and yellowhammers. In places, red squirrels can now cross the common without having to touch the ground.

Looking back as I stop for a breather, I can see all the way down into Swindale. The sunlight is reflecting on the floodplain wetlands either side of the beck as it winds its lazy way down the valley. The meadows are just starting to take on some colour, and I can make out a small group of sheep grazing in one of the woody pastures on the valley sides.

As I crest the ridge, I get into the wind-clipped mossy heath of the tops. Dwarf willow, cowberry and stiff sedge stud the mat of lichen and woolly fringe moss, while bearberry sprawls over the rocks. A wheatear chacks angrily at me and shows its white arse as it flies away. As the habitat lower down has become scrubbier and woodier, the wheatears have moved up the hill, and seem none the worse for it. I wonder if we'll have any dotterel, small waders from the plover family, up here this year. They don't breed every year, but it's become a much more regular occurrence since these moss heaths recovered.

I press on to the cairns on Artlecrag Pike and enjoy the uninterrupted views of Harter Fell, Kidsty Pike and High Street, and the intense satisfaction in knowing that they too all have plenty more life than when I first set eyes on them. I've spent most of my working years in this place. Shoulder to shoulder with friends and colleagues, we've coaxed nature back in, helped the landscape to function again. It still has a long way to go, but it's in a much better state than when I started, and that's all any of us can hope for.

As I stand and stretch, ready for the walk back down the hill, I'm briefly darkened by the massive shadow of a golden eagle. She circles and fixes an eye on me, head tilted, before flying off over the wild landscape that she rules again.

AFTERWORD

Since finishing this book, nothing has stood still.

Matthew and his young family secured a farm tenancy of their own. Their new place is only a few minutes away, so he's still involved, helping us out with hay cutting and gathering. David and his wife, Faith, have taken over the care of the Haweswater sheep, cattle and ponies, and are busily hatching plans for how to integrate them into a fully regenerative farming system.

There are other irons in the fire. Together with neighbouring landowners, we've secured a major grant to help nature across a swathe of eastern Cumbria. Over the course of five years, the funding will pay for the restoration of the missing tarn in Sale Pot, the return of water voles and marsh fritillary butterflies, create more jobs and support farmers who want to transition to nature friendly farming practices.

Designs for the renovation of one of the historic barns at Naddle Farm are taking shape, which will give us a better space to host education visits and provide offices for our growing Haweswater team.

It's not all rosy. Negotiations to secure a new stewardship agreement on Bampton Common on the west side of the reservoir are proving taxing. The commoners, landowners and government officers can't agree on how the land should best be managed.

It's been a turbulent time globally. A war in Europe broke out on the day the hardback version of this book was published. As well as being a tragedy for those caught up in it, the war sent shockwaves across the continent, disrupting energy and supply chains. Food security has risen up the agenda, threatening the stability of the bridges that were being built between farmers and environmentalists. Domestic government dramas haven't helped, with the environment seemingly being used as a political football, kicked in a different direction with every ministerial shuffle. This creates huge uncertainty for those of us working on the ground.

This, I suppose, is how it goes, and how it will continue. New opportunities, new challenges, adaptation, evolution. The work will never be done.

ACKNOWLEDGEMENTS

My first thanks should go to Haweswater itself, to its rocks, water and wild inhabitants, for the inspiration they have given me. Writing a book wasn't something I ever planned to do, and I'm still rather bemused by the fact that I've done so. I've simply responded to the landscape's insistence that there was a story here that needed to be told.

Like every farm, our work at Haweswater is a team effort, and there are many people who have contributed over the years. To my brilliant past and present RSPB colleagues: Hannah Bernie, Matthew and Dani Blair, Murray Craig, Jo Chamberlain, Heather Devey, Mal Edwards, Rhys Findlay-Robinson, David and Faith Garvey, Trev and Lucy Hughes, Bill Kenmir, Steve Kershaw, Ashley Lyons, David Morris, Vanessa Moss, Bea Normington, Annabel Rushton, Dave Shackleton, Anna Shiel, Richard Smith, Jim Wardill, Spike Webb and all our wonderful, dedicated volunteers – your energy and enthusiasm make coming to work every day pure pleasure. If you'd like to keep up to date with what we're doing, follow us on Twitter @wildhaweswater or visit wildhaweswater.co.uk.

Haweswater is one small part of the mighty RSPB family. There are too many colleagues that have chipped into the work at Haweswater to mention everybody. Kevin Cox and Alex Try deserve special personal thanks for championing my book from the start and believing that it might help to bolster our organization's mission in some small way. I hope that it does. The RSPB is incredible. If you aren't a member already, please join.

None of what we do at Haweswater would be possible without the close working relationship we have with United Utilities, and particularly John Gorst and the local catchment team. Our successes are your successes.

One of the best things about working in the Lakes is how much the lines between different organizations are starting to blur and how much support the many people working in conservation give to each other. Pete Barron at John Muir Trust, Gareth Browning and Victoria Lancaster at Forestry England, John Pring and Rachel Oakley at the National Trust, Pete Leeson at the Woodland Trust, Kevin Scott, Stephen Trotter and all at Cumbria Wildlife Trust, Olly Southgate at the Environment Agency, Simon Webb and Jean Johnstone at Natural England and many others are combining passion with profession to make a difference for nature. Thank you for sharing your knowledge and your accounts of the good and bad times that you've had while standing up for Cumbria's wildlife. The story I've told here is as much yours as it is mine.

Sincere thanks to the many Lake District farmers and advisors who have taken the time to show me around their land and to broaden my outlook on farming and the good that it can do. John Atkinson and Maria Benjamin at High Nibthwaite, Sam and Claire Beaumont at Gowbarrow Hall, Caroline Grindrod at Wilderculture, Can and Sam Hodgson at Glencoyne Farm, Jim Lowther, David and Jim Bliss at Lowther Estate, Richard Maxwell in Ennerdale, Andrea Meanwell at Low Borrowbridge, Paul and Nic Renison at Cannerheugh, Danny Teasdale at Ullswater CIC and Michael Wentworth Waites at Thornthwaite Hall. Thanks also to James Rebanks for the education and inspiration. Long may the ranks of nature-friendly farmers continue to swell.

Beyond the Lakes, my thanks go to Philip Ashmole and all at Borders Forest Trust, Charlie Burrell and Isabella Tree at the Knepp Estate, Helen Cole and all at NTS Ben Lawers, Duncan Halley and Dagmar Hagen at NINA in Norway, Stephen Westerberg and the team at RSPB Geltsdale. The lessons I learned from you and your places have had a huge influence on this book.

Thanks to Sara Barnard, Kevin Cox, Martin Fowlie, Martin Harper, Jean Johnstone and Alex Try, who read drafts of the book and provided invaluable advice. Any mistakes or missteps that remain are mine alone.

The description in chapter 15 of how plant life in the Pennines changed during the foot-and-mouth crises is reprinted with kind permission from Jeremy Roberts and Friends of the Lake District.

There would be no book at all if it weren't for the good fortune of meeting Katharine Norbury. I'm most grateful for your tweet that allowed us to get to know each other, and for making that vital introduction in Soho, where I was a fish out of water, and for all your support, encouragement and editorial input since.

The inestimable Patrick Walsh at PEW Literary has encouraged and supported me from day one and found me a happy home at Transworld. A total beginner like me couldn't have hoped for a more nurturing agent.

My editor Alex Christofi helped me to craft something that I never thought myself capable of. Richard Mason did a first-class job of copy-editing and Irene Martínez Costa used Alex Green's spectacular cover illustration to design a book that looks like Haweswater in paper form. Thanks also to Lilly

Cox, Catriona Hillerton, Sharika Teelwah, Susanna Wadeson, Katrina Whone, Sally Wray and everyone else at Transworld who helped bring my book into being.

Most of all, to my beautiful, precious family. To Becki, thank you for countless read-throughs and edits, boundless patience, reassurance, love and support; I'm incalculably lucky that you've chosen to share your life with me. To Elliot and Aphra – the nature-rich world that I'm striving for is for you to inhabit, please look after it.

BIBLIOGRAPHY

Addy, S., James Hutton Institute and International Union for Conservation of Nature (2016). *River Restoration and Biodiversity: Nature-Based Solutions for Restoring the Rivers of the UK and Republic of Ireland*. Aberdeen, Scotland: The James Hutton Institute.

Ardron, P. A. (1999). *Peat Cutting in Upland Britain, with Special Reference to the Peak District: Its Impact on Landscape, Archaeology, and Ecology*. PhD thesis, University of Sheffield. Available at: http://etheses.whiterose.ac.uk/6023/

Ashmole, M. and Ashmole, P., Borders Forest Trust and John Muir Trust (2009). *The Carrifran Wildwood Story: Ecological Restoration from the Grass Roots*. Jedburgh, Scotland: Borders Forest Trust.

Ashmole, M. and Ashmole, P. (eds) (2020). *A Journey in Landscape Restoration: Carrifran Wildwood and Beyond*. Dunbeath, Scotland: Whittles Publishing.

Austrheim, G., Speed, J. D. M., Evju, M., Hester, A., Holand, Ø., Loe, L. E., Martinsen, V., Mobæk, R., Mulder, J., Steen, H., Thompson, D. B. A. and Mysterud, A. (2016). 'Synergies and trade-offs between ecosystem services in an alpine ecosystem grazed by sheep – An experimental approach', *Basic and Applied Ecology*, 17(7), 596–608.

BBC, 'Switzerland's farmers become landscape gardeners', BBC News, 24 January 2018. Available at: https://www.bbc.co.uk/news/world-europe-42731932

Burns, F., Eaton, M. A., Gregory, R. D. et al. (2013). *State of Nature Report*. The State of Nature partnership.

Clark, C. and Scanlon, B. (2019). *Less is More: Improving Profitability and the Natural Environment in Hill and Other Marginal Farming Systems*. Available at: https://nt.global.ssl. fastly.net/documents/hill-farm-profitability-report-pdf.pdf

Clarke, R. (2014). *Two Hundred Years of Farming in Sutherland: The Story of my Family*. Isle of Lewis: The Islands Book Trust.

Cocker, M., Mabey, R., Gomersall, C., Elphick, J. and Macdonald, H. (2020). *Birds Britannica*. London: Chatto & Windus.

Cox, K., Groom, A., Jennings, K. and Mercer, I. (2018). 'National parks or natural parks: how can we have both?', *British Wildlife*, 30(2), 87–96.

Cumberland and Westmorland Herald (2015). 'Slurry leak killed thousands of fish in pristine tributary', *Cumberland and Westmorland Herald*, Penrith. Available at: https:// www.cwherald.com/a/archive/slurry-leak-killed-thousands-of-fish-in-pristine-tributary.479717.html

Cumbria County Council (2002). *Cumbria Foot and Mouth Disease Inquiry Report*. Available at: https://www.cumbria. gov.uk/eLibrary/Content/Internet/538/716/37826163827.pdf

Cumbria Wildlife Trust (2012). *Cumbria Wildlife Trust – High Fell – Wilson's Story – Farming Changes*. Available at: https://www.youtube.com/watch?v=l_oDXSf4JHU

Department for Business, Energy and Industrial Strategy (2020). *2018 UK Greenhouse Gas Emissions, Final figures*. Available at: https://assets.publishing.service.gov.uk/ government/uploads/system/uploads/attachment_data/ file/862887/2018_Final_greenhouse_gas_emissions_ statistical_release.pdf

Donald, M. B. (1994). *Elizabethan Copper: The History of the Company of Mines Royal, 1568–1605*. Ulverston, Cumbria: Red Earth Publications.

Edwards, M. (2017). *Sheep Farming on the Lake District Fells: Adapting to Change*. Available at: http://www.cumbria commoners.org.uk/files/mervyn_edwards_lake_district_ fell_shepherds_final_jan_2018.pdf

Eurostat (2019). *Agriculture Statistics – Family Farming in the EU – Statistics Explained*. Available at: https://ec.europa.eu/ eurostat/statistics-explained/index.php/Agriculture_ statistics_-_family_farming_in_the_EU

Gambles, R. (2013). *Lake District Place Names*. South Stainmore, Kirkby Stephen, Cumbria: Hayloft Publishing Ltd.

Gibbons, B. and Mabey, R. (1996). *Flora Britannica*. London: Sinclair-Stevenson.

Glover, J. (c. 1824). *Ullswater, Early Morning* (oil on canvas). Art Gallery of New South Wales, Australia.

Gooley, T. and Gower, N. (2017). *How to Read Water: Clues and Patterns from Puddles to the Sea*. London: Sceptre.

Gow, D. (2020). *Bringing Back the Beaver: The Story of One Man's Quest to Rewild Britain's Waterways*. White River Junction, Vermont: Chelsea Green Publishing.

Halley, D. (2015). 'History and harvesting of red deer and other cervids in Norway' (PowerPoint presentation). Available at: https://www.nina.no

Halley, D. (no date). 'Mountain birch (Betula pubescens tortuosa)' (PowerPoint presentation). Available at: https:// www.nina.no

Halliday, G. (1997). *A Flora of Cumbria*. Lancaster: Centre for North-West Regional Studies.

Hoffman, J. (2020). *Irreplaceable: The Fight to Save our Wild Places*. London: Penguin Books.

Jepson, P. and Blythe, C. (2020). *Rewilding: The Radical New Science of Ecological Recovery*. London: Icon Books Ltd.

Knapp, S. (2019). 'Are humans really blind to plants?', *PLANTS, PEOPLE, PLANET*, 1(3), 164–8.

Lake District National Park Authority (2013). *Tourism*. Available at: http://www.lakedistrict.gov.uk/learning/factstourism

Lake District National Park Partnership (2017). *Nomination of The English Lake District*. Available at: https://whc.unesco.org/document/155906

Lake District National Park Partnership (2018). *State of the Park 2018: Reporting on the Partnership's Plan 2015–2020*. Available at: https://www.lakedistrict.gov.uk/__data/assets/pdf_file/0018/151038/SOTP-Report-2018-V6-FINAL-02.05.19.docx.pdf

Leopold, A. (2020). *A Sand County Almanac: And Sketches Here and There*. London: Penguin Books.

Leopold, A. and Leopold, L. B. (1993). *Round River: From the Journals of Aldo Leopold*. Oxford: Oxford University Press.

Loren, B. (2013). *Animal, Mineral, Radical: Essays on Wildlife, Family, and Food*. Berkeley, California: Counterpoint.

Macdonald, B. (2019). *Rebirding: Rewilding Britain and its Birds*. Exeter, UK: Pelagic Publishing.

Macpherson, H. A. (1892). *A Vertebrate Fauna of Lakeland*. Edinburgh: David Douglas.

Maeght, J. L., Rewald, B. and Pierret, A. (2013). 'How to study deep roots – and why it matters', *Frontiers in Plant Science*, 4, 299.

Marrs, R. H., Lee, H., Blackbird, S., Connor, L., Girdwood, S. E., O'Connor, M., Smart, S. M., Rose, R. J., O'Reilly, J.

and Chiverrell, R. C. (2020). 'Release from sheep-grazing appears to put some heart back into upland vegetation: A comparison of nutritional properties of plant species in long-term grazing experiments', *Annals of Applied Biology*, 177(1), 152–62.

Mawdsley, T., Chappell, N. A. and Swallow, E. (2017). *Hydrological Change on Tebay Common Following Fencing and Tree Planting: A Preliminary Dataset*. Available at: http://www.es.lancs.ac.uk/people/nickc/TebayCommon_HydrologicalChange_REPORT_final.pdf

McCormick, T. (2018). *Lake District Fell Farming: Historical and Literary Perspectives, 1750–2017*. Carlisle: Bookcase.

Meanwell, A. (2015). 'As a shepherd, I know we have not "sheepwrecked" Britain's landscape', *Guardian*, 21 July. Available at: https://www.theguardian.com/commentisfree/2015/jul/21/farmers-sheep-lake-district-preserve-environmentalists

Meikle, R. D. and Gordon, V. (2001). *Willows and Poplars of Great Britain and Ireland*. London: Botanical Society of the British Isles.

Mitchell, W. R. (1993). *The Lost Village of Mardale*. Settle, North Yorkshire: Castleberg.

Molloy, D. (2011). *Wildlife at Work: The Economic Impact of White-Tailed Eagles on the Isle of Mull*. The RSPB, Sandy.

Monbiot, G. (2013). *Feral: Searching for Enchantment on the Frontiers of Rewilding*. London: Allen Lane.

Monbiot, G. (2017). 'The Lake District as a world heritage site? What a disaster that would be', *Guardian*, 9 May. Available at: https://www.theguardian.com/commentisfree/2017/may/09/lake-district-world-heritage-site-george-monbiot

National Farmers Union (2019). *Achieving NET ZERO. Farming's 2040 Goal.* Available at: https://www.nfuonline. com/nfu-online/business/regulation/achieving-net-zero-farmings-2040-goal

Natural England Lake District Team (2020). *Grazing Regimes for Nature Recovery: Experience from 25 Years of Agri-Environment Agreements in the Lake District's High Fells.* Available at: http://www.wildennerdale.co.uk/wordpress/ wp-content/uploads/2020/11/Grazing-Regimes-for-Nature-Recovery-7.7.20-.pdf

Nature Conservancy Council (1985). *Citation: Lamonby Verges and Fields.* Available at: https://designatedsites. naturalengland.org.uk/PDFsForWeb/Citation/1002257.pdf

Nilsen, E. B., Milner-Gulland, E. J., Schofield, L., Mysterud, A., Stenseth, N. C. and Coulson, T. (2007). 'Wolf reintroduction to Scotland: public attitudes and consequences for red deer management', *Proceedings of the Royal Society B: Biological Sciences*, 274 (1612), 995–1003.

OECD (2015). *Environment at a Glance 2015: OECD Indicators.* Paris: OECD Publishing. Available at: https:// doi.org/10.1787/9789264235199-en.

Omond, R. (1910). 'Large differences of temperature between the Ben Nevis and Fort-William observatories', *Transactions of the Royal Society of Edinburgh*, 44(2), 702–5.

Page, C. (1982). 'The history and spread of bracken in Britain', *Proceedings of the Royal Society of Edinburgh. Section B. Biological Sciences*, 81(1–2), 3–10.

Pakeman, R. J., Le Duc, M. G. and Marrs, R. H. (2000). 'Bracken distribution in Great Britain: strategies for its control and the sustainable management of marginal land', *Annals of Botany*, 85(2), 37–46.

Panda (2019). *Nature's Treasure Trove*. Available at: http://wwf.panda.org/knowledge_hub/where_we_work/alps/area/species2/

Parliamentary Office of Science and Technology (2009). *Deer Species and Populations in the UK*. Available at: https://www.parliament.uk/documents/post/postpn325.pdf

Pattison, I., Hardy, R. and Reaney, S. (2009). *Long-Term Changes in Flood Risk in the Eden Catchment, Cumbria: Links to Changes in Weather Types and Land Use*. Available at: https://www.researchgate.net/publication/252962689_Long_term_changes_in_flood_risk_in_the_Eden_Catchment_Cumbria_Links_to_changes_in_Weather_Types_and_Land_Use

Pinches, C. E., Gowing, D. J. G, Stevens, C. J., Fagan, K. and Brotherton, P. N. M. (2013). 'Natural England review of upland evidence – Upland Hay Meadows: what management regimes maintain the diversity of meadow flora and populations of breeding birds?', *Natural England Evidence Review,* no. 005. Available at: http://publications.naturalengland.org.uk/publication/5969921

Plantlife (2012). *Our Vanishing Flora – How Wild Flowers are Disappearing across Britain*. Salisbury: Plantlife.

Prospect (2020). *The State of Natural England 2020–21. A View from Prospect Trade Union*. Available at: https://library.prospect.org.uk/download/2020/01173

Ratcliffe, D. (2000). *In Search of Nature*. Leeds: Peregrine Books.

Ratcliffe, D. (2002). *Lakeland*. London: Collins.

Raven, J. E. and Walters, S. M. (1956). *Mountain Flowers*. London: Collins.

Rebanks, J. (2015). *The Shepherd's Life: A Tale of the Lake District*. London: Penguin Books.

Rebanks, J. (2021). *English Pastoral: An Inheritance*. London: Penguin Books.

Relph, J. (2014/15, Autumn/Winter). 'Looking Forward', *Federation of Cumbria Commoners Newsletter*. Available at: http://www.cumbriacommoners. org.uk/files/fcl_winter_2014_newsletter_0.pdf

Richardson, G. (2019). *Hows and Knotts: A Guide to Lakeland Views*. Faringdon: Redshank Books.

Roberts, F. J. (2010). *Pyramidal Bugle Ajuga pyramidalis in Kentmere Status in 2010*. Unpublished.

Roberts, J. (2010). 'The flowering of Cross Fell: montane vegetation and foot-and-mouth', *British Wildlife,* 21(3), 160–67.

Rose, F., O'Reilly, C., Smith, D. P. J. and Collings, M. (2006). *The Wild Flower Key: How to Identify Wild Flowers, Trees and Shrubs in Britain and Ireland*. London: Frederick Warne.

RSPB (2015). *Farming with Nature at RSPB Haweswater*. The RSPB, Sandy.

RSPB (2018). *Farming at Haweswater. An Economic Report 2013–2016*. The RSPB, Sandy.

Rural Business School at Duchy College (2018). *The Value of the Sheep Industry: North East, South West and North West Regions. A report by the Rural Business School at Duchy College on behalf of the National Farmers' Union*. Available at: https://www.nfuonline.com/assets/106083

Savills (2020). *Global Farmland Index. Spotlight Savills Research*. Available at: https://www.savills.co.uk/research_ articles/229130/304987-0

Savory, John (2016). 'Colonisation by woodland birds at Carrifran Wildwood: the story so far', *Scottish Birds*, 36, 135–49.

Schofield, L., Johnstone, J., Heritage, G. and Southgate, O. (2017). 'Rewilding in a Managed Landscape – The Swindale Beck Restoration Project', *In Practice* (March), 21–5.

Schofield, L., Teasdale, D., Hampson, D. and Ausden, M. (2020). 'Balancing culture and nature in the Lake District', *British Wildlife*, 31(4), 254–63.

Schwartz, J. D. (2014). 'Soil as carbon storehouse: new weapon in climate fight?', *Yale E360*, 4 March. Available at: https://e360.yale.edu/features/soil_as_carbon_storehouse_new_weapon_in_climate_fight

Scott, M. (2016). *Mountain Flowers*. London: Bloomsbury Natural History.

Shrubsole, G. (2019). *Who Owns England? How We Lost Our Green and Pleasant Land, and How to Take It Back*. London: HarperCollins Publishers.

Shrubsole, G. (2021). 'Life finds a way: in search of England's lost, forgotten rainforests', *Guardian*, 29 April. Available at: https://www.theguardian.com/environment/2021/apr/29/life-finds-a-way-in-search-of-englands-lost-forgotten-rainforests

Silvis, H. and Voskuilen, M. (2018). *Agricultural Land Prices in the EU in 2016*. Available at: https://pdfs.semanticscholar.org/deb2/6998f9335c460159df71633533368113ec00e.pdf

Stewart, R. (2013). 'Rory challenges environmentalists to protect small upland farms', *Rory Stewart*, 23 October. Available at: http://www.rorystewart.co.uk/rory-challenges-environmentalists-to-protect-small-upland-farms

Stewart, R. (2014). 'Swindale – What future for our landscape?', *Rory Stewart*, 31 January. Available at: http://www.rorystewart.co.uk/swindale-future-landscape

Stewart, R. (2016). *The Marches*. London: Jonathan Cape.

Stott, M., Callion, J. C., Kinley, I., Raven, C. and Roberts, J. (2002). *The Breeding Birds of Cumbria: A Tetrad Atlas 1997–2001*. Keswick: Cumbria Bird Club.

Stroh, P., Walker, K., Smith, S., Jefferson, R. and Blackstock, T. (2019). *Grassland Plants of the British and Irish Lowlands : Ecology, Threats and Management*. Bristol: Botanical Society of Britain and Ireland.

Thórsson, A. T., Pálsson, S., Sigurgeirsson, A. and Anamthawat-Jónsson, K. (2007). 'Morphological variation among Betula nana (diploid), B. pubescens (tetraploid) and their triploid hybrids in Iceland', *Annals of botany*, 99(6), 1183–93.

Treanor, J., Kollewe, J. and Farrell, S. (2015). 'Storm Desmond damage across Cumbria estimated at £500m', *Guardian*, 8 December. Available at: https://www.theguardian.com/uk-news/2015/dec/08/storm-desmond-damage-cumbria-estimated-500m

Tree, I. (2018). *Wilding: The Return of Nature to a British Farm*. London: Picador.

Varley, M. and Friends of The Lake District (2003). *Flora of the Fells: Celebrating Cumbria's Mountain Landscapes*. Kendal: Friends of The Lake District.

Wainwright, A. (2019). *The North Western Fells: Wainwright's Walking Guide to the Lake District. Book 6*, ed. Clive Hutchby. London: White Lion Publishing.

Wandersee, J. H. and Schussler, E. E. (1999). 'Preventing plant blindness', *The American Biology Teacher*, 61, 84–6.

Wandersee, J. H. and Schussler, E. E. (2001). 'Toward a theory of plant blindness', *Plant Science Bulletin*, 47, 2–8.

Watts, S. H., Griffith, A. and Mackinlay, L. (2019). 'Grazing exclusion and vegetation change in an upland grassland with patches of tall herbs', *Applied Vegetation Science*, 22 (3), 383–93.

Whaley, D. (2006). *A Dictionary of Lake District Place-Names*. Nottingham: English Place-Name Society.

Wikipedia (2020). *Cumbrian Toponymy*. Available at: https://en.wikipedia.org/wiki/Cumbrian_toponymy

Wikipedia (2020). *Philip Ashmole*. Available at: https://en.wikipedia.org/wiki/Philip_Ashmole

Wilson, E. O. (1996). 'The environmental ethic', *3 Hastings West Northwest Journal of Environmental Law and Policy*, 327. Available at: https://repository.uchastings.edu/hastings_environmental_law_journal/vol3/iss2/12

Wordsworth, D. (1987). *The Grasmere journal: the revised complete text*. London: Michael Joseph.

Wordsworth, W. (2004). *Selected Poems*, ed. Stephen Gill. London: Penguin Books.

INDEX

Lee Schofield is site manager at RSPB Haweswater in the Lake District, a landscape-scale nature reserve incorporating working farms. *Wild Fell* is his first book.